Nuclear Energy Leadership

NUCLEAR ENERGY LEADERSHIP

Lessons Learned From U.S. Operators

Mary Jo Rogers, PhD

PennWell®

Disclaimer

The recommendations, advice, descriptions, and the methods in this book are presented solely for educational purposes. The author and publisher assume no liability whatsoever for any loss or damage that results from the use of any of the material in this book. Use of the material in this book is solely at the risk of the user.

PennWell Corporation
1421 South Sheridan Road
Tulsa, Oklahoma 74112-6600 USA

800.752.9764
+1.918.831.9421
sales@pennwell.com
www.pennwellbooks.com
www.pennwell.com

Marketing Manager: Amanda Alvarez
National Account Executive: Barbara McGee

Director: Mary McGee
Managing Editor: Stephen Hill
Production Manager: Sheila Brock
Production Editor: Tony Quinn
Book Designer: Susan E. Ormston

Library of Congress Cataloging-in-Publication Data Available on Request

Rogers, Mary Jo
Nuclear Energy Leadership: Lessons Learned from U.S. Operators
ISBN 1-59370-245-0
ISBN13 978-1-59370-245-8
CIP DATA if available

Printed in the United States of America

1 2 3 4 5 17 16 15 14 13

For Rich and Julius

Contents

Preface

With the disastrous events at Japan's Fukushima Daiichi station on March 11, 2011, the safety and viability of nuclear power again came to the forefront of the public dialogue. The record earthquake and tsunami destroyed towns and took thousands of lives, flooded the Daiichi nuclear units, and wiped out power supplies and the backup diesel generators designed to cool the reactor and the stored used nuclear fuel. If good can come from terrible loss, then Fukushima may ultimately bring clarity to the global risks of nuclear energy—and how to minimize them.

While fossil-derived energy has fueled extraordinary growth in much of the world's economy, the full cost of burning fossil fuels is increasingly realized in terms of its environmental impact and the loss of human life. When the Macondo well operated by BP and Transocean started blowing millions of gallons of oil into the Gulf of Mexico on April 20, 2010, the explosion killed 11 men, destroyed countless Gulf creatures and plant life, and severely damaged the regional coastal economy. Although this disaster was one of many deadly ocean drilling blowouts, refinery explosions, and coal mining accidents in recent years, it garnered the public's attention regarding the costs of a fossil energy–hungry economy.

Many countries have relied on nuclear power as a key part of their energy portfolios and, like the United States, plan to continue using nuclear generation into the future. Other countries are in the process of building a number of new nuclear power plants. Nuclear power has become more attractive over the years because of its negligible carbon impact. Nuclear power is also locally generated, and so it does not need to be imported. But nuclear power plants are very expensive to build and extremely challenging to run well. Fortunately, there are many nuclear operators who have figured out how to run nuclear power plants with exceptional safety and reliability.

This book is not a polemic on the complicated history of nuclear energy or on the policies shaping the future of nuclear power. Nor is it a technical manual on how to run a nuclear power plant. *Nuclear Energy Leadership: Lessons Learned from U.S. Operators* is a discussion of how to lead and manage power plant organizations in order to achieve and maintain the highest possible levels of safety and production simultaneously. This book is based on the hundreds of observations, thousands of interviews and assessments, and essential lessons learned over a period of nearly 15 years of immersion in the nuclear power industry. It is a tribute to those leaders and managers who make the commitment every day with hope that others, whether running nuclear power plants or other high-risk operations, will learn from the effort.

Acknowledgments

I am indebted to the committed men and women who work hard every day to keep nuclear energy safe, reliable, and carbon free. I have met many admirable individuals in the nuclear power business. For those who took their time to share their experiences and viewpoints with me, I am very grateful.

Thanks to the industry leaders and professionals who let me interview them in depth, including Preston Swafford (TVA), Bill Levis (PSEG), Bill Mohl (Entergy), Jeff Merrifield (Shaw), Mark Schiavoni (APS), Joe Pollock (NEI), Rick Purcell (INPO), Tim Rausch (PPL), Ed Halpin (PG&E), Jim Lynch (INPO), Donna Jacobs (Entergy), Bob Hovey (Exelon), Eric Olson (Entergy), Ron Detwiler (TVA), Bonnie Stahl (INPO), James Mertink (STP), Laura Brieden (STP), Les Guthrie, Richard Cole, Ed McVey (Exelon), and Derek Hawes (TVA).

I especially thank the remarkable Oliver Kingsley (Exelon, retired) for teaching many of us about operating excellent, safe nuclear power plants and developing leadership capability at the same time. I am very grateful to the relentless high standards of Chris Crane (Exelon), who encouraged me to pursue my interests in nuclear energy leadership, team, and organizational development. And thanks to Jack Davis (Detroit Edison, retired) for his passion for leadership and his confidence in my work. My apologies to the great leaders and managers I have yet been unable to interview.

I also want to thank Steve Hill at PennWell, who was supportive and persistent. Thanks to my business partner, John Philbin at Strategic Talent Solutions, for understanding my absences. I appreciate the kindness of Penni Conner, who was inspirational and instrumental.

I thank my parents, now gone, for their encouragement of my endeavors, and I thank my sisters for their emotional support. Finally, thanks to my husband Rich, whom I admire and adore, and my son Julius, who is my pride and joy.

CHAPTER 1 State of U.S. Nuclear Energy

The United States' commercial nuclear power industry has steadily improved its safety, production, and financial performance over the past 30 years to become the safest, most reliable, and lowest-cost electricity generator.[1] U.S. nuclear power has made an industry-wide transformation since March of 1979 when a series of events at the Three Mile Island (TMI) nuclear power plant near Middletown, Pennsylvania, resulted in a partial reactor core meltdown and a public much more concerned about nuclear plant safety. The disastrous incidents at the Daiichi nuclear power plant in Fukushima, Japan, in March 2011 subsequent to the Great East Japan Earthquake and Tsunami have awakened the public's anxieties about nuclear power. Yet the Fukushima disaster comes at a time when U.S. nuclear power plants have become the safest industrial working environments worldwide and overall plant performance is at an all-time high:

- Nuclear stations' total industrial safety accident rate has been at or below 0.21 accidents per 200,000 worker-hours since 2001 (fig. 1–1).[2]
- Nuclear generating *unit capability*, the percentage of time the plant is online and producing electricity, has been at or above 91% since 2001 (fig. 1–2).[3]
- Nuclear electricity production costs were 2.14¢ per kilowatt-hour (kWh) in 2010 and have hovered around 2¢/kWh beginning in 2001 (fig. 1–3).[4]

Japan's record earthquake and tsunami and the resulting accident at the Fukushima Daiichi nuclear power plant had a powerful impact on the views of nuclear energy worldwide. Before elaborating on the merits of effective nuclear energy operators and how to learn from the best practices of the industry, it is important to review the events surrounding the Fukushima disaster and discuss the impact on the U.S. nuclear energy industry.

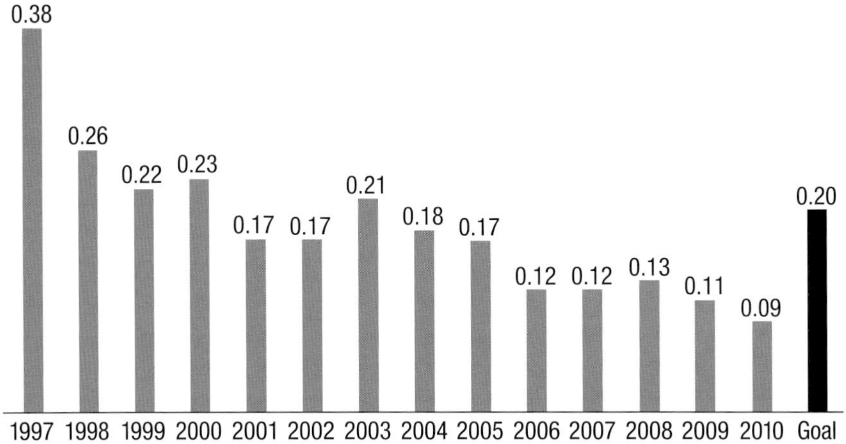

Fig. 1–1. U.S. nuclear power industrial safety accident rate: one-year industry values. ISAR = Number of accidents resulting in lost work, restricted work or fatalities per 200,000 worker hours. Note: Starting in 2008, data includes supplemental personnel. (Source: *Nuclear News*. 2011. "INPO's U.S. Reactor Fleet Performance Indicators for 2010 Show Mixed Results." June, 30–31.)

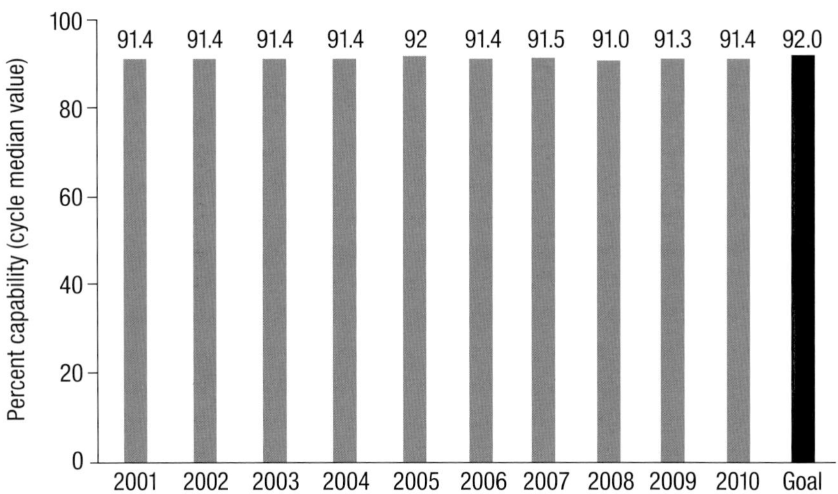

Fig. 1–2. U.S. nuclear power industry unit capability. (Source: *Nuclear News*. 2011. "INPO's U.S. Reactor Fleet Performance Indicators for 2010 Show Mixed Results." June, 30–31.)

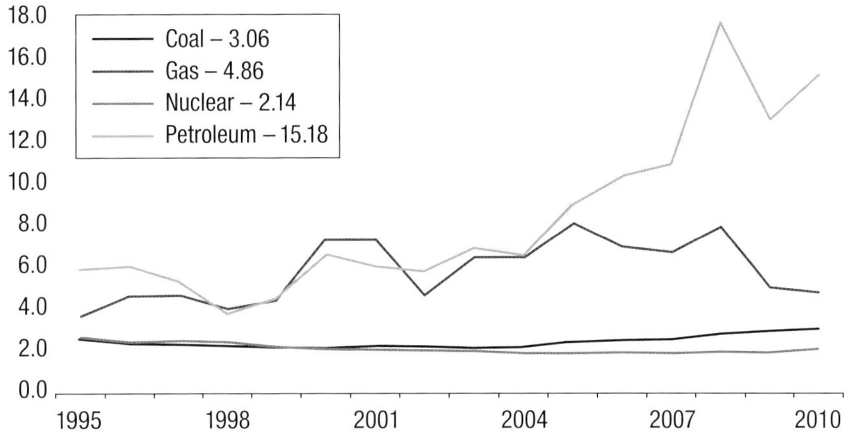

Fig. 1–3. U.S. electricity production costs, 1995–2010, in 2010 cents per kilowatt-hour. Production Costs = Operations and Maintenance Costs + Fuel Costs. Production costs do not include indirect costs and are based on FERC Form 1 filings submitted by regulated utilities. Production costs are modeled for utilities that are not regulated. (Source: Nuclear Energy Institute. 2011. *U.S. Electricity Production Costs 1995–2010.* http://www.nei.org/resourcesandstats/documentlibrary/reliableandaffordableenergy/graphicsandcharts/uselectricityproductioncosts/.)

The Disaster at Fukushima

On March 11, 2011, when a record 9.0 magnitude earthquake struck off the Pacific coast of the Tohoku region in Japan, 11 reactors at four sites automatically shut down as designed. The earthquake caused terrible devastation in the area, and off-site power was entirely lost. Emergency diesel generators provided power to the Daiichi units until 40 minutes later, when a large tsunami wave overwhelmed the site's protection barrier and caused the loss of emergency power to four of the six units at the Daiichi station, located approximately 150 miles north of Tokyo.[5] Without power to pump cooling water to the plants' reactors and spent fuel pools, some of the nuclear fuel began to melt, and hydrogen gas explosions occurred inside the units, resulting in radiation releases. Extraordinary efforts were underway to bring water in to cool the reactors, and much work was done to stabilize the units and spent fuel cooling pools over a period of days and weeks.[6]

The massive earthquake and tsunami caused widespread destruction across northeastern Japan, where 15,391 lives were lost and 8,171 people were missing. Under extreme conditions, the nuclear power plant's operator, Tokyo Electric Power Company (TEPCO), worked diligently to protect workers and nearby communities, and no fatalities have been attributed to radiation exposure.[7] The radiation releases from the Daiichi units led Japanese authorities to order evacuations, and with subsequent readings outside the plant, officials alerted people not to consume tap water, milk, fish, and beef from the immediate area.[8] Although trace amounts of radiation from Japan were found in British Columbia and Washington State days later, the amount was small: 100,000 times lower than natural background radiation.[9]

By December 2011 the Daiichi units were brought to a state of safe shutdown as recovery efforts at the site and community continued. The Japanese government has taken many proactive steps to protect the public from potential health impacts. The area around the plant is undergoing extensive monitoring, remediation, and decontamination of the land and structures.[10]

U.S. nuclear industry impact

The impact of the Fukushima disaster on the U.S. nuclear industry has been intense and instantaneous and is likely to be enduring. The U.S. Nuclear Regulatory Commission (NRC) alerted its incident response staff, and U.S. nuclear companies, as well as the NRC, sent experts and resources to Japan to support emergency response and containment efforts within days of the incident. U.S. nuclear power plant operators took actions without delay to verify the capabilities of their safety and emergency response systems, and seismic, flooding, and loss-of-power mitigation planning. NRC inspectors were directed to make immediate independent assessments of each plant's preparedness. The NRC launched a comprehensive safety review, starting with the near-term (90-day) task force, within two weeks of the event to evaluate initial lessons learned from Fukushima.[11] The NRC also initiated a long-term task force, which will have the opportunity for greater in-depth technical study of the events of Fukushima and potential implications for U.S. nuclear plant safety.

In July 2011 the near-term task force that conducted the initial investigation of the Fukushima incident and U.S. plant preparedness reported,

> The current regulatory approach, and more importantly, the resultant plant capabilities allow the Task Force to conclude that a sequence of events like the Fukushima accident is unlikely to occur in the United States and some appropriate mitigation measures have been implemented, reducing the likelihood of core damage and radiological releases. Therefore, continued operation and continued licensing activities do not pose an imminent risk to public health and safety.

The task force noted that some of what has strengthened the U.S. nuclear plants' capabilities resulted from the numerous emergency and security upgrades the industry made after the attacks on the World Trade Center and the Pentagon on September 11, 2001. The task force went on to recommend additional enhancements to safety and emergency preparedness for the commission's consideration.[12] The complete long-term impact of Fukushima on the nuclear power industry remains to be seen.

Public concern and questions about the safety of U.S. nuclear plants were ubiquitous in the weeks following the Fukushima disaster. Industry, company, and site leaders responded with multiple communications locally and nationally about steps that have been taken to ensure the safety and security of the plants in local communities and the U.S. nuclear power fleet. Although polls showed a decline in support for nuclear power immediately after Fukushima, six months later a majority (67%) of Americans rated U.S. nuclear power plant safety "high."[13]

Critics of the nuclear power industry remain vigilant, however, and they have pointed out potential problems at some of the U.S. nuclear plants in the wake of the Fukushima disaster with regard to emergency preparedness for simultaneous problems at multiple units, availability of backup power for spent fuel pools, and flooding.[14] In order to speak to such concerns and to keep the public informed of its response to the Fukushima events, the NRC has released multiple communications on the steps plant operators and the NRC have taken to address each of these areas. As the problems surrounding the catastrophic events at the Daiichi plants become better known and understood, the nuclear power industry has publicly committed to adopting the lessons learned into current and future plant operations.[15]

Future of U.S. Nuclear Power

Commercial nuclear power continues to be a significant source of electricity in the United States, providing approximately 20% of the nation's electricity every year since 1990. As a result of its high reliability and low production cost, nuclear power is more intensive than all other sources. While nuclear power is only one-tenth of the total U.S. electricity generation capacity, nuclear plants generate one-fifth of the electricity used—because most nuclear plants run continuously to provide baseline megawatts to the grid and to get the best use out of their nuclear fuel (fig. 1–4).[16]

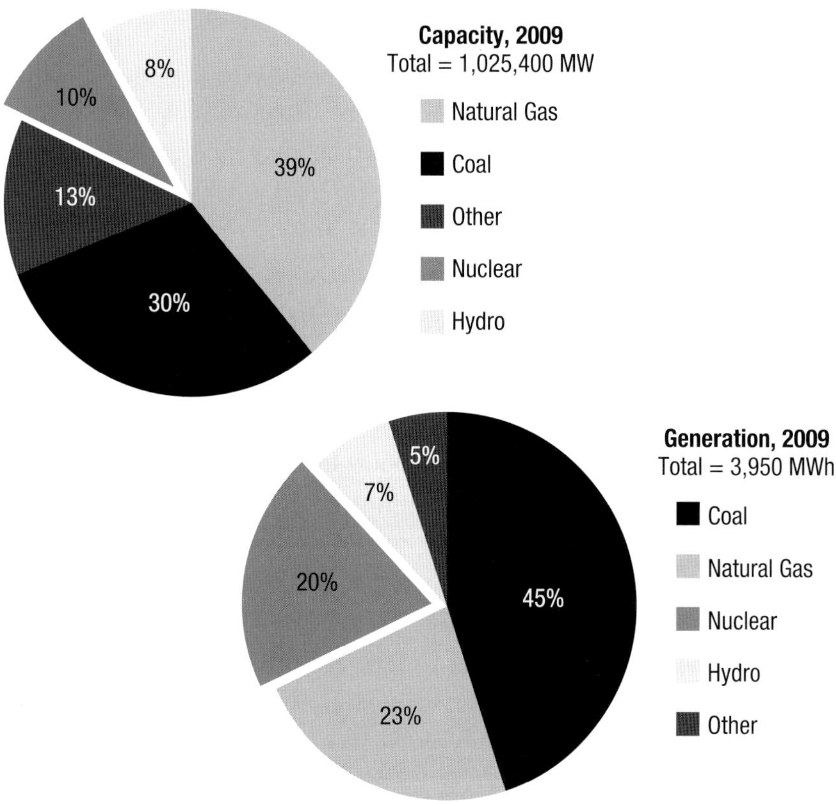

Fig. 1–4. U.S. electricity capacity and generation, 2009. Nuclear plants use more of their capacity to generate electricity than other plants. (Source: U.S. Energy Information Administration. 2012. *Electric Power Monthly*, July, table 1.1.)

In the decade before the Fukushima disaster, nuclear power experts had anticipated a surge in the building of new nuclear units across the United States to meet demand for reliable, low-carbon electricity. The much-anticipated surge in new construction, sometimes referred to as the nuclear renaissance, slowed substantially prior to the events at Fukushima as a result of a precipitous drop in natural gas prices and electricity rates. Extensive U.S. shale natural gas reserves became accessible via the emergence of hydraulic fracturing, or *fracking*. Although some people have raised environmental concerns, including potential contamination of underground water due to fracking, such questions have not yet significantly affected the pace of extraction or natural gas prices. Continued low natural gas prices have made nuclear power construction much less appealing to potential investors. Nuclear power's low operating and fuel costs are offset by high costs of construction, which are dramatically lower for natural gas plant construction. The recession in the United States and lower-than-expected demand for electricity also slowed new nuclear power plant construction.[17]

Despite these setbacks, there are many indications that nuclear power will continue to be a fundamental part of the U.S. energy portfolio, but with a more gradual expansion. Although Germany's Chancellor Angela Merkel announced the phasing out of Germany's nuclear power program after the Fukushima disaster and public protest, official U.S. backing for safe nuclear power has not wavered. President Barack Obama and Energy Secretary Steven Chu reaffirmed the safety of the U.S. nuclear fleet and endorsed the continuation of the process for licensing and building new reactors. Released shortly following the Fukushima disaster, the President's energy plan included continued support for safe nuclear power, stating that "our nuclear power plants have undergone exhaustive study to ensure that they can safely withstand a number of extreme contingencies."[18]

Many nuclear generation construction projects are moving forward, including the building of Vogtle Units 3 and 4 in Waynesboro, Georgia, by the Southern Company—the first new construction and operation license (COL) granted by the NRC in 33 years.[19] As of December 31, 2011, the NRC had 17 active COL applications for new nuclear power plant reactors from nine different companies. Other new license applications have been voluntarily put on hold by six applicants.[20] In addition to new nuclear construction, expansion is seen in *power uprates* (which can allow higher power output from the original reactor hardware), license extension, and the return to work on partially built units. For example, construction has restarted at Watts Bar Unit 2, a Tennessee Valley Authority (TVA) site in eastern Tennessee. Multiple plants continue to pursue and obtain license renewal, such as Palo Verde Units 1, 2, and 3 owned by Arizona Public Service Company. One of many power uprate approvals includes Point Beach Units 1 and 2 owned by NextEra Energy, which were approved for a 17% thermal power increase for each unit.[21]

The performance and vitality of the U.S. nuclear power fleet is important to the industry worldwide. Although some U.S. plants do lag behind the industry in various areas, they are few and they are not allowed to fall far behind (see chap. 2). The United States is a nuclear power leader in many ways, producing 27% of the world's nuclear-generated electricity. France is the next-largest nuclear power generator, at 17% (fig. 1–5). The United States has increased its capacity factor 3.5% over the past 10 years to a rolling average of approximately 90%.[22] The U.S. NRC's annual Regulatory Information Conference in March 2012 reported more than 3,000 participants from 35 different countries. The conference included multiple programs on the U.S. and international responses to Fukushima.[23]

The nuclear power industry has retained the confidence of U.S. government officials and continually works to earn the public's trust. Without achieving and maintaining high levels of performance in safety and reliability, it is questionable whether the U.S. nuclear power industry would be in its current position. Moreover, the nuclear power industry's significant, sustained improvements in safety and generation would have been unlikely without transformational changes and a relentless commitment to safety.

Learning from U.S. Nuclear Power Operators

U.S. nuclear power leaders and organizations have evolved immensely in order to achieve and maintain high levels of performance in safety and production in what is arguably the most closely regulated industry. Research shows that there are leadership and organizational characteristics that have been pivotal to the industry's improvements. These characteristics are the behaviors and practices that the industry has adopted over time that have led

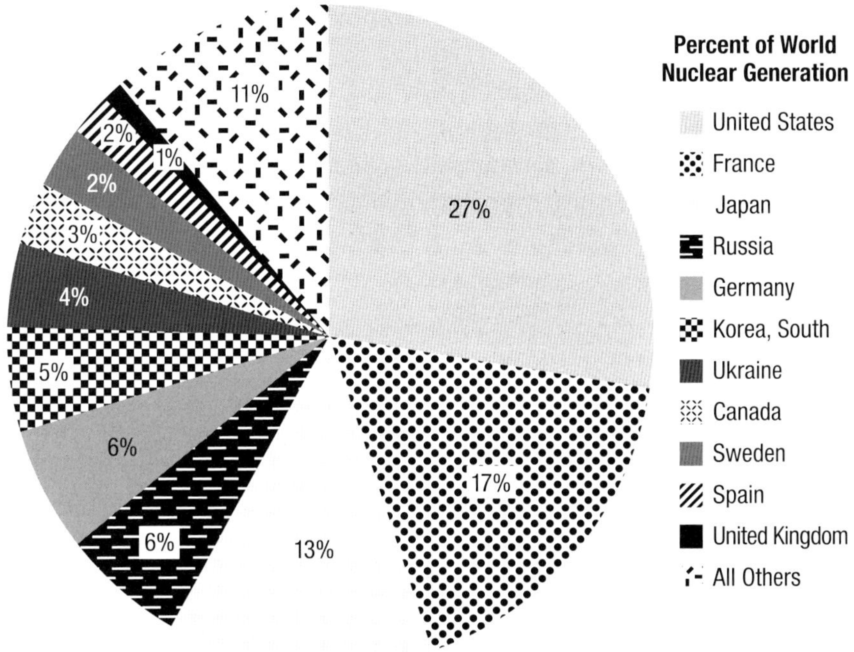

Fig. 1–5. Gross nuclear electric power as a percent of world nuclear generation, 2009. Because of independent rounding, the figures may not add up to the total percentage. The country's short-form name is used. (Source: International Atomic Energy Association, Power Reactor Information System, as of May 6, 2010.)

to sustained levels of high performance, and they comprise the significant lessons learned that are captured in the following chapters. The lessons learned from U.S. nuclear power are divided into three parts:

- Industry-wide advances in self-regulation and safety
- Organizational changes and best practices that are executed at the sites
- Individual nuclear leader and leadership team behaviors

Although these practices are drawn from the nuclear power industry, there are aspects that can be applied to high-risk operations and heavy industry where managers strive to simultaneously improve safety and production. In addition, industries can benefit from learning more about these effective approaches in anticipation of increased regulatory pressure and oversight.

To support the application of nuclear practices outside of the nuclear industry, the conclusion of each chapter is devoted to experiences with and suggestions for transferring nuclear lessons learned to other industrial environments. The last section of the book includes tools that can be used to assess and intervene at the team and organizational levels and a review of nuclear and related electricity generation organizations.

References

1 U.S. Energy Information Administration. 2003. "Nuclear Power: 12 Percent of America's Capacity, 20 Percent of the Electricity." May. http://www.eia.gov/cneaf/nuclear/page/analysis/solution4.pdf.

2 *Nuclear News*. 2011. "INPO's U.S. Reactor Fleet Performance Indicators for 2010 Show Mixed Results." June, 30–31.

3 Ibid.

4 Nuclear Energy Institute. 2011. *U.S. Electricity Production Costs 1995–2010*. http://www.nei.org/resourcesandstats/documentlibrary/reliableandaffordableenergy/graphicsandcharts/uselectricityproductioncosts/.

5 Miller, Charles, Amy Cubbage, Daniel Dorman, Jack Grobe, Gary Holahan, and Nathan Sanfilippo. 2011. *Recommendations for Enhancing Reactor Safety in the 21st Century: The Near-Term Task Force Review of Insights from the Fukushima Dai-Ichi Accident*. Washington, DC: U.S. Nuclear Regulatory Commission. July 12. pbadupws.nrc.gov/docs/ML1118/ML111861807.pdf.

6 U.S. Nuclear Regulatory Commission. 2011. *Fact Sheet on Summary of Japan Events in March 2011 and NRC Response*. September. http://www.nrc.gov/reading-rm/doc-collections/fact-sheets/fs-japan-events.html (updated March 29, 2012).

7 International Atomic Energy Agency. 2011. *IAEA International Fact Finding Expert Mission of the Fukushima Dai-Ichi NPP Accident Following the Great East Japan Earthquake and Tsunami: Report to the IAEA Member States*. May 24–June 2. http://www-pub.iaea.org/mtcd/meetings/pdfplus/2011/cn200/documentation/cn200_final-fukushima-mission_report.pdf.

8 Osnos, Evan. 2011. "The Fallout. Seven Months Later: Japan's Nuclear Predicament." *New Yorker*, October 17, 46–61.

9 Burgess, Joe, Amanda Cox, Sergio Peçanha, Amy Schoenfeld, and Archie Tse. 2011. "Assessing the Radiation Danger, Near and Far." *New York Times*, April 3, 12.

10 Nuclear Energy Institute. 2012. "Japan: Fukushima Recovery Efforts." March 20. http://safetyfirst.nei.org/japan/.

11 Jaczko, Gregory B. 2012. "Moving Forward for Safety." Remarks by the Honorable Gregory B. Jaczko, Chairman U.S. NRC at the Regulatory Information Conference. U.S. Nuclear Regulatory Commission. March 13. http://pbadupws.nrc.gov/docs/ML1208/ML120800466.pdf.

12 Miller, Charles, et al. 2011. vii–viii.

13 Nuclear Energy Institute. 2011. "Americans' Support for Nuclear Energy Holds at Majority Level 6 Months after Japan Accident." News release, October 3. http://www.nei.org/newsandevents/newsreleases/americans-support-for-nuclear-energy-holds-at-majority-level-6-months-after-japan-accident/.

14 Wald, Matthew L. 2011. "Disaster Plan Problems Found at U.S. Nuclear Plants." *New York Times*, May 13. http://www.nytimes.com/2011/05/13/business/energy-environment/13nuke.html.

15 Jaczko, Gregory B. 2011. "A Regulator's Perspective on the Challenges of Today and Tomorrow." *NRC News*. U.S. Nuclear Regulatory Commission, Remarks presented at the Nuclear Energy Assembly, Charlotte, NC, May 11.

16 U.S. Energy Information Administration. 2011. "What Is the Status of the U.S. Nuclear Industry?" April 22. http://www.eia.gov/energy_in_brief/nuclear_industry.cfm (updated June 11, 2012).

17 Maize, Kennedy, Charles Butcher, and Robert Peltier. 2011. "Global Gas Glut." *Power*, September, 46–55.

18 *Nuclear News*. 2011. "U.S. Power Reactors to Be Examined for Vulnerabilities." May, 22–28.

19 "Vogtle Nuclear Expansion Gets First Federal Approval in 33 Years." 2012. POWERnews, February 15. http://www.powermag.com/POWERnews/4390.html?hq_e=el&hq_m=2384453&hq_l=4&hq_v=402b5b2e5f.

20 *Nuclear News*. 2012. "New Power Reactor Projects in the United States." March, 88.

21 *Nuclear News*. 2011. "Maximizing the Assets: A Status Report on License Renewal and Power Uprates." September, 24.

22 U.S. Nuclear Regulatory Commission. 2010. *Information Digest, 2010–2011*. NUREG-1350, vol. 22, August, 23–26.

23 ———. 2012. Regulatory Information Conference Program, March 13–15.

CHAPTER 2 Self-Regulation

The significant improvements in the U.S. nuclear power industry are partly attributable to self-regulation achieved through the work of the Institute of Nuclear Power Operations (INPO).[1] INPO was formed by utility executives in 1979 after the contained partial core meltdown of Unit 2 at Three Mile Island (TMI) station. Together with industry leaders, INPO gradually shaped commercial nuclear power into a highly self-critical, cooperative culture that is continually striving for excellence. Standards of excellence, industry cooperation, and information sharing are foundational to the effectiveness of self-regulation in U.S. nuclear power.

There is renewed interest in the successful self-regulation that has occurred in nuclear power. The report by the National Commission on the BP Deepwater Horizon Oil Spill refers to civilian nuclear power plant operators as providing a model for effective self-policing.[2] INPO's CEO, James Ellis, delivered an invited presentation about INPO to the oil spill commission. Generally, however, INPO has kept a low profile over the years in order to protect the confidentiality of its members while promoting openness within the group. Unfortunately, little is known about how INPO works outside of nuclear power industry management. Other industries that simultaneously pursue improvements in safety and production could benefit from learning some of the lessons from nuclear energy's self-regulation.

INPO is not the only self-regulatory body that helps the nuclear power industry continually improve. There are also nuclear safety review boards, quality assurance and oversight organizations, and corporate functions that play a role. These groups are under the auspices of the nuclear companies themselves, but serve an independent oversight purpose and provide observations and feedback to nuclear management and leadership on a regular basis. It could be argued that the success of nuclear self-regulation also owes something to the effectiveness of the U.S. Nuclear Regulatory Commission (NRC).

This chapter reviews the primary functions of INPO and the role of leadership, cooperation, and information sharing in nuclear self-regulation. Company-supported external and internal oversight organizations are examined. This chapter also presents the potential impact of the NRC on self-regulation and the nuclear industry's performance. The chapter concludes with a discussion of transferring nuclear self-regulation lessons to other industries, with a special focus on oil and gas.

Institute of Nuclear Power Operations (INPO)

In the years leading up to TMI, there was little perceived need for cooperation and industry-wide self-regulation among the nuclear power utilities. The construction and operation of nuclear power plants were conducted relatively independently by the utility companies just as they had run other parts of their businesses. The TMI event was the first major nuclear power accident. Although there was not a harmful release of radiation and core damage was completely contained by the reactor vessel, public concern and mistrust of nuclear power intensified, partly because the situation was handled so poorly by government and company officials at the time.[3]

After TMI, nuclear utility leaders united, recognizing that if any one plant had a catastrophic event, it would be devastating for the entire industry. The leaders acknowledged that they needed to work together to raise standards and plant performance. Utility CEOs formed INPO even before the president's commission that investigated the TMI accident released its final report.[4] The commission recommended that the industry "set and police its own standards of excellence to ensure the effective management and safe operation of nuclear power plants."[5] In order to do so, industry leaders recognized that they needed to work together, share best practices, and raise standards and performance across the board.

INPO today is a privately incorporated, independent industry organization made up of about 400 employees (staffing levels have essentially remained the same over time). INPO has a budget of $99 million, which is drawn almost entirely from dues from its members, including all of the 26 companies and utilities that operate 104 nuclear power reactors in the United States. INPO's mission, to promote excellence in the operation of commercial nuclear power plants, is achieved through the four cornerstone programs (see table 2–1).[6]

Table 2–1. Institute of Nuclear Power Operations' four cornerstone programs

Evaluation	Training
Every operating site receives a plant-wide evaluation and assessment from INPO at least every two years. Evaluation teams identify strengths and areas for improvement based on standard criteria. Evaluation teams include industry peers from other sites. Industry peer evaluators share standards and best practices with on-site management. Evaluation and assessments play a crucial role in elevating safety and operating performance across the nuclear power industry.	INPO maintains two training roles: • The evaluation and accreditation of specific training programs in operations and maintenance and technical areas. • The provision of nonaccredited training courses for supervisors, managers, and plant managers, as well as seminars for executives. Training courses have class sponsors who mentor the class and share their experiences. Training programs provide an informal mechanism for information sharing and socializing nuclear professionals on the values of cooperation, self-development, and learning.
Analysis	**Assistance**
INPO continually collects extensive plant performance data from all operators monthly and when incidents occur. The analysis organization analyzes the data for issues and trends. Trends, issues, and recommendations are communicated to the institute's membership. INPO issues mandatory follow-up to significant events and trends.	Assistance teams visit plants to provide additional insights and suggestions regarding specific areas for improvement. The assistance group includes senior representatives, who are experienced professionals who serve as liaisons between the site and INPO. Every site is assigned a senior representative. Poorer performing plants receive additional help and oversight by way of special assistance visits.

Evaluation

INPO accomplishes its self-regulatory function primarily through its evaluation cornerstone. Every two years, each plant in the United States receives a thorough evaluation of its safety and operational performance in the following areas: operations, maintenance, engineering, work management, radiation protection, chemistry, security, emergency preparedness, training, performance improvement, and organizational effectiveness, among others. The evaluation is a specific review of plant programs, procedures, employee behaviors, and results. The evaluation team uses INPO's performance objectives and criteria (POs and Cs) to assess performance in each of the areas. The POs and Cs serve as the industry's de facto standards. INPO develops standards that are distinct from the federal legal requirements of the NRC and other regulatory agencies (e.g., state and environmental), although they are consistent with them. INPO's official aspirational standard is excellence in operational reliability and safety. Thus, the POs and Cs are typically higher than the legal minimum requirement and change over time as the institute recognizes safer and better ways of operating nuclear power plants.

The evaluation team consists of INPO personnel who are experts in each of the areas, as well as loaned employees and industry peer evaluators. Loaned employees, or loanees, are nuclear utility employees who work full-time for INPO for an agreed-upon time period. Industry peer evaluators essentially volunteer to spend approximately three weeks as part of the INPO evaluation team for another company's plant evaluation. In 2009, INPO's combined evaluation teams conducted 34 plant evaluations across the country.[7]

Evaluators are painstaking about providing feedback to the site and the utility owner's leaders. In addition to daily debriefings at the plant about the evaluators' hypotheses and findings, there is a full overview at the conclusion of the evaluation period followed by the presentation of the report to site and company leaders. The team identifies strengths and areas for improvement (AFIs). The plant's leadership takes the feedback in a final report and writes a response on how they are going to improve the problem areas. Most plants today expend a great deal of effort and resources to fix their AFIs.

Informally, peer evaluators and loanees will offer ideas based on how they operate at their home stations and take lessons learned back to their own companies. Plant personnel also will be told what sites to benchmark based on who is strong in the areas where the plant is weak. By using nuclear personnel from other site organizations on evaluations, INPO has facilitated site managers becoming apt to call or visit one another to gather information on best practices when they need help, thereby fostering industry cooperation and information sharing.[8]

Training

Improved training for operators was one of the main recommendations of the president's commission that studied the TMI accident.[9] INPO has taken on two training roles. In one role, they provide the evaluation and accreditation of site training programs for the operations and maintenance and technical areas through an organization called the National Academy for Nuclear Training. INPO establishes the training program requirements for the industry, but the companies themselves develop and execute the programs. INPO's accredited programs provide standards and guidelines for site training programs, such as for initial license training and requalification training for plant operators. (Under federal regulation, the NRC tests and grants licenses to individuals to operate the plants.) Accredited programs are evaluated every four years by the academy board, which includes industry members in addition to academy professionals.

Secondly, INPO provides training directly, for example, for supervisors and new plant managers. INPO's training programs offer many opportunities for leaders, managers, and supervisors to meet people from other plants and exchange helpful information. INPO also runs leadership development seminars and courses for the new nuclear employee, the nuclear supervisor, manager, plant manager, site VP, and even the board members of a nuclear utility.[10]

Most courses have class mentors, or senior nuclear leaders who sponsor the class by sharing experiences and answering questions. Not only do INPO's accredited programs and leadership development courses expose nuclear leaders to a common set of standards and operating experience, they also perpetuate common cultural values. Such shared values include the importance of industry cooperation, continuous improvement, self-development, and learning.

Analysis

The analysis organization at INPO collects industry data, analyzes it for issues and trends, and communicates results and recommendations with the industry. The Plant Information Center (PIC) collects all performance data from the sites on a continuous basis. Since 2009, all sites have had access to the PIC data for all stations, thereby facilitating comparisons and benchmarking. INPO also screens about 2,000 plant events per year looking for significant issues to share with all its members so they can prevent similar events from occurring elsewhere.[11] The analysis organization formally issues notices on significant events, requiring all plants to take specific actions, which are later evaluated for compliance.

In one sense, INPO equalizes its members' opportunities to perform by providing every plant access to operating experience, as well as information on how to prevent and solve problems. The constant flow of information, with elevated communications when needed, also has helped to raise performance levels for the entire fleet of plants across the nation. According to Richard Purcell, INPO senior VP of industry performance improvement, industry safety and reliability performance indicators are near historic highs, while the number of low-performing plants is at its lowest. "Nonetheless," Purcell says, "we have to be vigilant and respond to early signs of the erosion of fundamentals."[12]

Assistance

The assistance cornerstone refers to INPO's commitment to provide extra help to struggling plants and to any plant with a specific need that asks for assistance. Special focus teams visit poorer performing plants, assessing and providing guidance in particular areas. Assistance teams will spend a week at a plant and assess an area such as equipment reliability or industrial safety. The teams include members from other plants that are proficient in the problem areas so they can share their ideas and experiences with the struggling plants.

Part of the assistance program includes the senior representative group. Each nuclear station is assigned a senior representative who is an INPO employee with plant experience. The senior representative's task is to bridge communications, share best practices, and monitor plant performance between INPO evaluations. The senior representative also facilitates use of assistance visits to help close gaps. In general, assistance activities are designed to keep higher performing plants continuously improving and lower performing plants from deviating too far from industry standards. For the latter, the frequency of assistance visits and communications with the sites' senior leaders help them turn around performance.[13]

Leadership Support for Cooperation and Information Sharing

The nuclear industry's cooperative approach to sharing information, best practices, and even resources is a powerful factor in the effectiveness of self-regulation and subsequently, the industry's sustained performance improvements. The culture of cooperation developed in the nuclear industry allows technical problems to be addressed, with shared solutions and lessons learned about equipment issues. Combined with the distribution of operational experience, this culture continues to have the effect of raising performance across the entire industry.

Example

A striking example of industry cooperation can be seen in how the chairman of one industry group, the Reactivity Control Review Committee (RCRC), helped the River Bend Station operations department with their reactivity control challenges in 2009. River Bend is a single-unit boiling water reactor (BWR) owned and operated by Entergy and situated approximately 25 miles north of Baton Rouge near St. Francisville, Louisiana. The RCRC is part of the BWR Owners' Group. Prior to 2009, the plant experienced a number of reactivity management issues that led to unplanned power reductions and shutdowns to deal with problems related to the reactor fuel itself and to fix systems that can potentially impact the core. (*Reactivity management* refers to the systematic monitoring and controlling of conditions and activities that can impact the core.)[14] River Bend and other U.S. stations monitor reactivity continuously and are prepared at all times to reduce power on the unit or take it off-line to address potential problems.

River Bend's operations department reactor engineer sought assistance from the RCRC and Edward McVey, the committee's chairman. The RCRC had been developing guidelines for excellence in reactivity management. McVey immediately offered to help the site and traveled to River Bend for a week to evaluate their reactivity management program and offer recommendations.

Ed McVey's full-time job is manager of reactor engineering oversight for Exelon, which has the nation's largest fleet of nuclear power plants (Entergy has the second largest fleet). McVey's work on the RCRC and his trip to assist River Bend were fully supported by the leaders above him.[15] Although Entergy and Exelon are considered competitors in that they both have plants that are part of unregulated electricity markets, on the nuclear generation side of the business, they are able to support one another in ways that improve overall industry performance.

According to Eric Olson, River Bend VP who was general manager of plant operations at the time, the site took action to implement McVey's suggestions, which included adding processes to improve controls, training people on their roles in reactivity management, and putting in place a reactivity management program that was in line with the RCRC's best practices.[16] After they made these changes, River Bend's reactivity management performance improved remarkably. When River Bend completed their INPO evaluation in 2010, they received high marks for their reactivity control plans. INPO noted that the site had not experienced any human performance reactivity errors during the two-year evaluation cycle.

Executive backing

Industry leaders advance cooperative culture in how they demonstrate support for the work of INPO and other groups. Executive support for INPO was not as strong in the past as it is currently. Joseph Rees's seminal book, *Hostages of Each Other*, analyzes the formation of INPO and its evolving influence on the commercial nuclear power industry. INPO has become a highly credible and influential organization whose impact, however, is interwoven with the power of peer pressure. Nuclear industry executives exert pressure on one another to conform to the cooperative norm, to stay engaged, and participate in self-regulatory activities.[17]

Although nuclear companies vary somewhat in their level of industry involvement, most actively participate in INPO offerings as well as other industry groups, such as the BWR Owners' Group and the RCRC. In addition to the RCRC discussed previously, the nuclear industry's cooperative culture is represented in numerous other industry groups. These groups meet throughout each year on various technical and nontechnical topics, where members share information including technical solutions (see app. C).

Nuclear sites regularly send people to travel with INPO as industry peers on evaluation teams. In addition, they allow employees to travel to other companies' sites to benchmark, and welcome others to examine best practices at their own plants. Industry leaders serve on INPO's board, advisory council, and expert panels on key topics, such as Leadership Fundamentals to Achieve and Sustain Excellent Station Performance.[18] Notably, member utility CEOs are expected to attend INPO's annual CEO conference without fail, and attendance is typically 100%.

Richard Purcell, INPO's senior VP of industry performance improvement, stated that the industry's willingness to help is impressive. "Company leaders factor into their budgets time and money to send people out on assistance and evaluation visits as well as for trips to INPO to collaborate and find out more about what others are doing best."[19] Moreover, INPO tracks the participation rates of its members and can identify sites that are trending toward low participation or potential isolation. Those sites are nudged to send leaders to a particular INPO activity and are reminded of the importance of industry involvement and the power of sharing information and best practices.

Company-Sponsored Self-Regulatory Functions

Nuclear companies have additional ways in which they obtain relatively independent feedback and opportunities to self-correct and improve. The three primary methods are nuclear safety review boards, quality assurance/oversight organizations, and corporate oversight organizations. While these groups vary in their degree of independence and in how they are leveraged across the nuclear industry, they serve an important self-regulatory function.

Quality assurance/oversight organizations

Nuclear sites are required by law to have internal quality assurance (QA) programs.[20] These programs include regular audits and inspections throughout the plant to monitor compliance with legal requirements pertaining to the design basis of the plant, operations, and maintenance, as well as other areas. In addition to the QA department, some companies have an assessment group that is not required by regulation, yet regularly conducts observations and assessments of site programs and personnel behaviors. These QA and assessment programs typically report up through the chief nuclear officer directly to keep them relatively independent of the line organization.

Exelon's nuclear oversight (NOS) organization is comprised of both the audit group and an assessment group. Exelon's assessment organization has systematic processes that codify how to conduct the duties of the assessor and NOS manager, in addition to the procedures for the required audit group. Robert Hovey served as Exelon's VP of nuclear oversight for five years, and, as part of Exelon Nuclear Partners, he is currently on loan in Japan with the Fukushima stabilization and recovery team. Many stations, from the United States and internationally, have visited Exelon to benchmark its NOS organization, as it is recognized as highly effective in helping Exelon "find problems before they find us." Hovey states that in order for the NOS organization to be value-added, its people need to find the right issues and have fortitude and leadership courage when challenging line leaders. "If NOS is focused on petty issues, leaders won't care." Additionally, "The single most important thing to NOS success is the support of the CNO and senior executives. The way you earn that support is by having the information, having it right, and being timely so that you can arrest the performance decline before most people can even see it."[21]

Nuclear safety review boards (NSRBs)

Each nuclear site is required to have an external review group, referred to as an NSRB, safety review council, or other names. The typical NSRB is comprised of approximately five to seven industry executives external to the company. Some of these executives are retired and serve on multiple nuclear boards and so bring a breadth of knowledge to the role. Each external member is dedicated to a particular site function, such as operations, maintenance, or engineering, for example. The remainder of the NSRB is made up of the company's top oversight and regulatory executives, such as the VP of nuclear oversight and director of licensing or VP of regulatory assurance. The NSRB meets one to four times annually on-site, spending the first day making observations, interviewing employees, and reviewing site data. The second day is the formal meeting of the NSRB, during which external board members present their findings and recommendations. Site leadership is present to discuss the feedback from the board.

The report submitted by the NSRB chairman is part of the site's formal quality assurance quarterly report. It is a technical requirement for nuclear stations to have an external review group that files a report with the quality assurance manager. Because the requirement is broadly written, however, sites have a good deal of flexibility in how they utilize the NSRB. In order to get the most out of the NSRB, which costs companies consultation fees paid to board members, sites have to make sure that the chairman and the members have high standards and that they are willing to provide critical feedback and challenge the site when necessary. To benefit, the sites have to take action on NSRB feedback.[22]

Corporate oversight organizations

Both the single-site nuclear company and the fleet have corporate oversight organizations, albeit to varying degrees. Although single sites may sometimes appear top heavy because they have multiple positions at the executive level (e.g., in engineering, QA, operations, or training), these executives also need to carry out various oversight functions. In addition to corporate oversight, nuclear fleets typically have corporate functional area managers that support the sites but have important oversight duties.

Corporate oversight is relatively less independent as compared to the NSRB, QA and assessment, and INPO. Still, corporate organizations are in a position to evaluate and identify underlying gaps between expectations and execution, share best practices, and coach site managers on how to improve. Because site personnel typically do not report to the corporate oversight leader, the oversight personnel need to have strong influence skills and the ability to build effective working relationships with site leadership. Corporate managers also need to have the mettle to escalate issues when site leaders remain unresponsive. In turn, for corporate functional area managers to be effective, senior executives need to support them when they do raise issues.

Nuclear Regulatory Commission (NRC)

Brief history

The NRC has been focused on its mission "to license and regulate the nation's civilian use of byproduct, source, and special nuclear materials in order to protect public health and safety, promote the common defense and security, and protect the environment" since it first began operation as an independent agency in 1975. In addition to the regulation of commercial nuclear power plants, the NRC regulates the uses of nuclear materials in medical, industrial, and academic settings, nuclear fuel production, and the transportation, storage, and disposal of nuclear waste. The NRC plays a key role in nuclear nonproliferation, as it licenses and regulates the import and export of radioactive materials.[23]

The NRC's predecessor, the Atomic Energy Commission (AEC), was formed by the Atomic Energy Act of 1946 to manage nuclear energy programs in the United States. The AEC essentially had dual roles of both developing and regulating nuclear power. A number of controversies over time led to the AEC's eventual reconstitution that divided up these responsibilities. The Energy Reorganization Act of 1974 established the NRC as an independent regulatory agency, separate from any promotional or developmental functions.[24] Nuclear energy development duties were taken on by the U.S. Energy Research and Development Administration, which was later subsumed by the Department of Energy. In 1994, various agencies combined to form the Nuclear Energy Institute (NEI) to perform public relations activities and represent the nuclear power industry on legislative issues.[25]

Over the years the NRC has made significant enhancements affecting commercial nuclear power operators, including changes in response to the accident at TMI in 1979. For instance, prior to TMI, the NRC concentrated on technical issues and risks to safety pertaining to plant equipment. The operator errors made during the TMI event brought about greater attention to human factors in plant operational problems. The failure to properly share previous operating experience that could have prevented the accident led to formalized operating experience programs. Other changes included greater requirements for operator training and licensing, increased research, and the stationing of two NRC resident inspectors at every site.[26]

There are multiple instances throughout the history of the NRC where the agency methodically evaluated industry issues and related criticism and has made regulatory enhancements. Two relatively recent examples include changes related to nuclear safety culture and fatigue management requirements.[27] Of course, not all NRC regulatory enhancements have been perceived as necessary improvements by all parties. Yet some nuclear energy industry critics see the agency as too accommodating to nuclear power plant owners.[28]

Potential NRC impact

When looking at the safety results of the nuclear power industry, improvements are seen in multiple areas. At an NEI conference in May 2011, the chairman of the NRC, Gregory Jaczko, reported that there were no statistically significant adverse trends in overall industry performance. "On an industry-wide basis, the performance indicators that we trend remained significantly improved compared to 10 or 15 years ago." Jaczko went on to say that they cannot be complacent, that there remain areas for concern, and that the industry must learn from the events at Fukushima.[29]

A review of the safety performance indicators for the U.S. nuclear fleet over the years shows that improvements can be seen in many areas, such as in significant events (fig. 2–1), safety system failures (fig. 2–2), unplanned automatic scrams (fig. 2–3), and forced outage rate (fig. 2–4). These performance indicators are consistent with others trends, such as in industrial safety and capacity factor (ch. 1), demonstrating broad and sustained improvements in the safety and reliability of U.S. nuclear plants over the past 10–15 years.

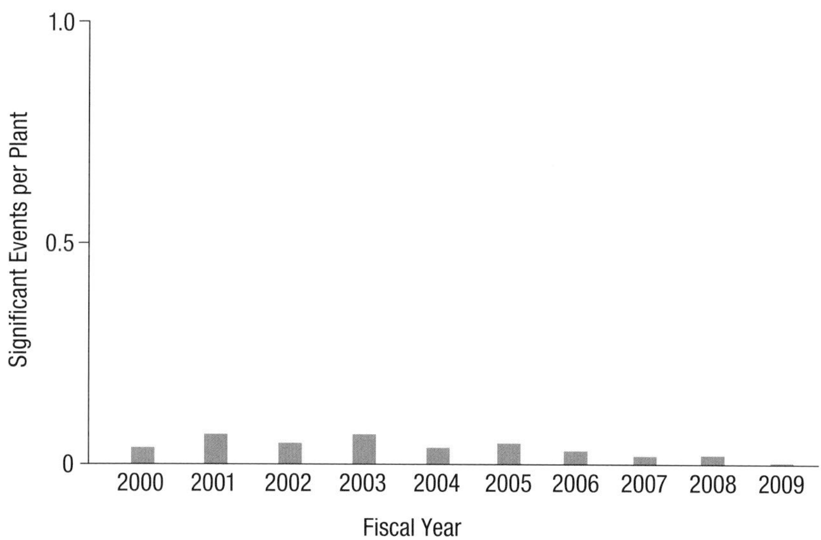

Fig. 2–1. U.S. nuclear power industry significant events. Significant events are events that meet specific NRC criteria, including degradation of safety equipment, a reactor scram with complications, or an unexpected response to a sudden degradation of fuel or pressure boundaries. The NRC staff identifies significant events through detailed screening and evaluation of operating experiences.

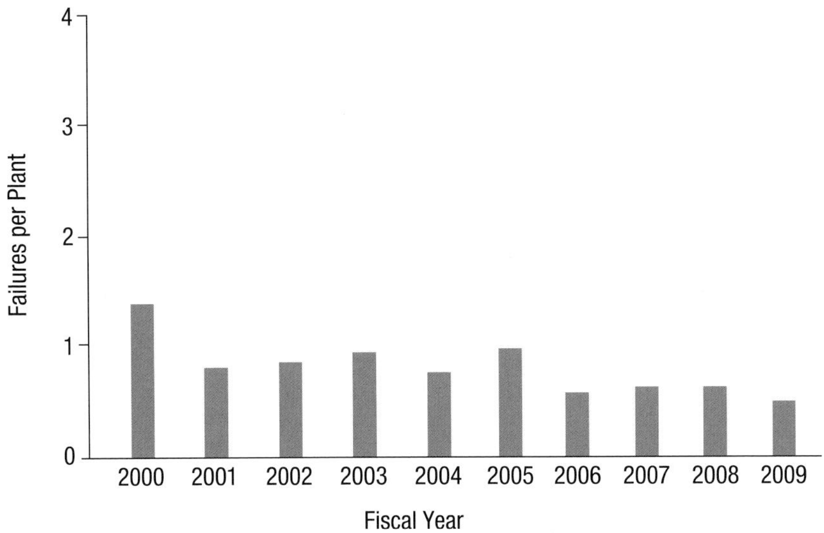

Fig. 2–2. U.S. nuclear power industry safety system failures. Safety system failures are any actual failures, events, or conditions that could prevent a system from performing its required safety functions.

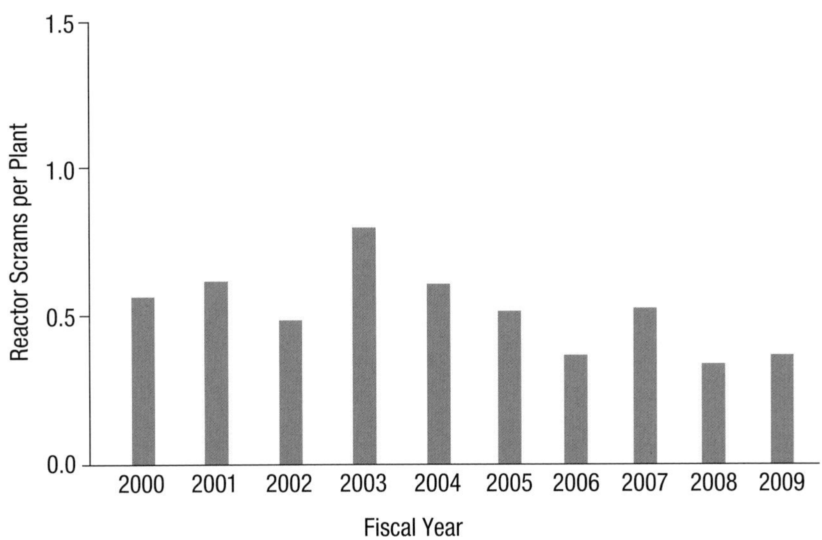

Fig. 2–3. U.S. nuclear power industry automatic scrams while critical. A reactor is said to be "critical" when it achieves a self-sustaining nuclear chain reaction such as when the reactor is operating. The sudden shutting down of a nuclear reactor by rapid insertion of control rods, either automatically or manually by the reactor operator, is referred to as a "scram." This indicator measures the number of unplanned automatic scrams that occurred while the reactor was critical.

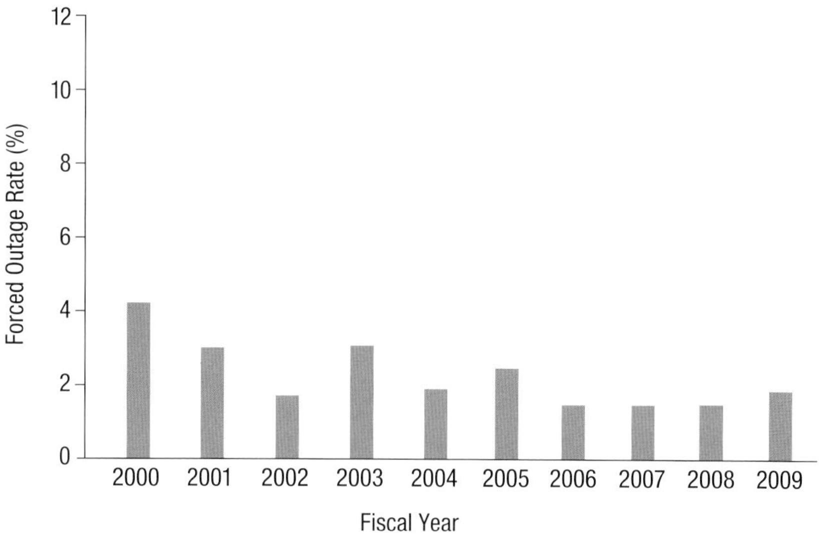

Fig. 2–4. U.S. nuclear power industry forced outage rate. The forced outage rate is the number of hours that the plant is unable to operate (forced outage hours) divided by the sum of the hours that the plant is generating and transmitting electricity (unit service hours) and the hours that the plant is unable to operate (forced outage hours).

During this time period of wide-ranging improvements, the NRC has been an active regulator. Present purposes do not allow for a review of NRC activities over the past decade. In sum, however, there are multiple examples of NRC leadership, involvement, and action in every area: rulemaking, guidance development, communications, licensing, decommissioning, inspection, enforcement conferences, investigations, events assessments, and adjudication.[30]

It would be difficult, if not impossible, to separate the effects of the work of the NRC from that of INPO, industry leaders, managers, the workforce itself, technological advances, and other factors in the overall improvement of nuclear power plant performance. One can imagine, however, the effect of a weak national regulator by contrast. James Lynch, VP of assistance at INPO, summarizes the relationship between the nuclear power industry and the NRC well: "The nuclear industry benefits from a strong regulator that is effective and fair. It is in the industry's best interests to have valuable oversight and the NRC is an important part of that oversight."[31]

International Nuclear Power Agencies

The International Atomic Energy Agency (IAEA) was created in 1957 by the United Nations to promote high levels of safety regarding the peaceful use of nuclear materials internationally. The IAEA is an independent intergovernmental agency that inspects and verifies nonproliferation agreements and works to ensure the protection of human health and the environment. In reaction to the Fukushima disaster, IAEA Director General Yukiya Amano proposed a framework to strengthen IAEA safety standards and reviews, focusing on safety margins against earthquakes, tsunamis, and floods.[32]

The World Association of Nuclear Operators (WANO) was formed in 1989 in response to the nuclear power plant disaster at the Chernobyl units in the former U.S.S.R., in an area that is now the Ukraine. Every power reactor owner worldwide participates in WANO, but WANO's influence can be limited by members' willingness to obtain critical feedback. In concert with the IAEA, WANO is also currently undergoing some self-examination and evaluation of how to increase its effectiveness. WANO is considering how to replicate the strong influence on plant owners that INPO achieves through critical and confidential plant assessments, for example.[33]

Applying Nuclear Self-Regulation Lessons

An informal poll of nuclear energy, utility, and oil and gas executives regarding the transfer of nuclear lessons learned to other industries revealed mixed results. Many responded in the negative, reflecting that the combined safety, regulatory, and risk environments of nuclear power are so unique as to make its practices inapplicable to other operations. Those who do see the benefit, such as William Levis, the COO of Public Service Electric and Gas (PSEG), tend to have had the experience of effectively transferring nuclear practices to nonnuclear operations and therefore readily see how nuclear concepts can be used.[34] By targeting a specific area, PricewaterhouseCoopers LLP also has drawn on the successful practices of the nuclear power industry (and others) regarding managing contractors conducting high-impact activities. The results are to be applied to organizations involved in high-impact activities that can have potentially catastrophic consequences.[35]

Others see the value of nuclear power experience in light of tragic events that may have been avoided and are aware of the track record of U.S. nuclear power plant safety. Approximately one year following the explosion of the Deepwater Horizon oil rig and the Macondo well blowout, Robert Dudley, CEO of BP, stated that the oil and gas industry does need to adopt lessons learned from the nuclear industry given its vast improvements in safety performance since the TMI accident. Speaking at the annual energy executive conference in Houston, Dudley praised the commercial nuclear energy industry for its ability to learn from the accident and make industry-wide advances in safety culture.[36]

Fundamentals of nuclear self-regulation

The oil spill commission recommended that the oil and gas industry draw on the nuclear power industry's experiences with regard to self-regulation. INPO CEO James Ellis presented to the commission five key factors that have enabled INPO and self-regulation to be effective in commercial nuclear power: chief executive officer engagement, nuclear power safety focus, support from the nuclear power industry, accountability, and independence.[37]

CEO engagement. Working directly with CEOs has been critical to achieving industry responsiveness to INPO's mission and various initiatives. INPO's board of directors is comprised of utility CEOs, who receive personal feedback from INPO on plant evaluation results.

Safety focus. INPO's mission, "To promote the highest levels of safety and reliability—to promote excellence—in the operation of commercial nuclear power plants," is vital to INPO's ability to raise nuclear power safety performance. INPO sets standards that are not based on regulatory compliance, but on finding the best and safest ways of operating. Since its founding, the institute has resisted any efforts to expand or dilute its mission.

Support from the industry. Commercial nuclear power plant leaders accept that they will undergo intrusive, comprehensive evaluations by INPO on a regular basis. INPO's programs and evaluations were initially resisted by some in the industry early in INPO history. Credibility was built over time as challenges and concerns were addressed. Currently INPO enjoys broad industry support and involvement, including participation in advisory groups, task forces, and loaned employee programs, among other activities.

Accountability. Accountability comes in various forms in this self-regulatory body. Making plant evaluations more critical and assigning a performance rating, while keeping reports confidential and limited in distribution, have made plant evaluations more effective. INPO will also put pressure on company leadership when plants have chronically low performance or are unresponsive to concerns when raised, or both. The nuclear power industry also has a collective insurance company, Nuclear Electric Insurance Limited (NEIL). Nuclear power plants are obligated to carry insurance by NEIL, which requires INPO membership and uses INPO assessment ratings to help determine premiums.

Independence. INPO has been careful over the years to maintain its independence while being a useful resource for the nuclear power industry. One way in which it protects its independence is by clearly distinguishing between its evaluation role and other more collaborative activities. INPO is deliberate about selecting particular team members for each plant evaluation. Additionally, INPO works in a complementary, constructive fashion with the NRC, although it remains independent of it.[38]

Initial steps by the oil and gas industry

The oil spill commission recommended improved regulatory oversight as well as better self-regulatory capabilities, particularly with regard to safety culture. One of the first actions taken since the BP disaster was the splitting up of the government's Minerals Management Services (MMS) to better protect its regulatory role. The MMS was renamed the Bureau of Ocean Energy Management, Regulation and Enforcement (BOEMRE) and divided into three groups in order to separate regulatory and enforcement responsibilities from leasing and revenue collections. The commission also recommended that the oil and gas industry form a new, independent, private entity responsible for building a definitive safety culture across the industry. It was suggested that this private organization would be able to serve as a liaison to government agencies and provide increased self-policing, but would need to guard against compromising on safety standards.[39]

In response to the disaster and commission report, the oil and gas industry established the Center for Offshore Safety, with a 22-member governing board of operators, drilling contractors, service and supply contractors, and trade association representatives. The center was created by the American

Petroleum Institute (API), contrary to the National Commission on the BP Deepwater Horizon Oil Spill's recommendation to establish a safety body independent of the industry group. Charlie Williams, Shell's former chief scientist of well engineering and production, played a key role in creating the new Center for Offshore Safety and was recently named its executive director. Williams has 40 years of energy industry experience and currently serves on the Department of the Interior's Offshore Energy Safety Advisory Committee. Williams will lead the center in one of its important activities, which is to conduct independent third-party audits to help ensure safe operating practices.[40]

The oil and gas industry may be seen as reluctant participants in any industry-wide self-regulatory activities. In an invited presentation to oil and gas executives at an industry conference, energy leaders discussed concerns about competitiveness and the need to protect proprietary information, which are viewed as primary obstacles to any self-regulatory activities resembling INPO.[41] Oil and gas leaders can yet take advantage of INPO experiences. Zack Pate, one of INPO's early CEOs, told the oil spill commission, "The largest barrier in the early days was the skepticism of a lot of members and the plant managers and the people working in the plant who have a very challenging job. We had to win the support and keep the support of the CEOs. Eventually if the CEOs support INPO programs, the guys down in the organization come along. Then it was our challenge to make them believe they should come along and build our credibility."[42]

References

1 Rees, Joseph V. 1994. *Hostages of Each Other*. Chicago: University of Chicago Press.

2 National Commission on the BP Deepwater Horizon Oil Spill and Offshore Drilling. 2011. *Deep Water: The Gulf Oil Disaster and the Future of Offshore Drilling*. Report to the President. January, 12. http://www.oilspillcommission.gov/sites/default/files/documents/DEEPWATER_ReporttothePresident_FINAL.pdf.

3 Rees, Joseph V. 1994.

4 Ellis, James O. 2010. "The Role of the Institute of Nuclear Power Operations in Self-Regulation of the Commercial Nuclear Power

Industry." Remarks before the National Commission on the BP Deepwater Horizon Oil Spill and Offshore Drilling. August 25. Nuclear Energy Institute. Public policy. http://www.nei.org/publicpolicy/congressionaltestimony/august-25-2010/.

5 President's Commission on the Accident at Three Mile Island. 1979. *Report of the President's Commission on the Accident at Three Mile Island. The Need for Change: The Legacy of TMI*. Washington, DC: President's Commission on the Accident at Three Mile Island. October, 68. http://www.threemileisland.org/downloads/188.pdf.

6 Ellis, James O. 2010.

7 Institute of Nuclear Power Operations. 2009. *Nuclear Safety: Setting the Global Standard*. Annual Report. Atlanta, GA: INPO.

8 Purcell, Richard. 2010. Interview with author. November 24.

9 President's Commission on the Accident at Three Mile Island. 1979.

10 Institute of Nuclear Power Operations. 2009.

11 Ibid.

12 Purcell, Richard. 2010. Interview with author.

13 Ibid.

14 Reactivity Controls Review Committee (RCRC). 2009. *Guidelines for Excellence. Section 5.0: Monitoring of Reactivity Management Issues*. Wilmington, NC: RCRC. July.

15 McVey, Edward. 2011. Interview with author. January 31.

16 Olson, Eric. 2011. Interview with author. October 21.

17 Rees, Joseph V. 1994.

18 Institute of Nuclear Power Operations (INPO). 2007. *Leadership Fundamentals to Achieve and Sustain Excellent Station Performance*. Atlanta, GA: INPO. September.

19 Purcell, Richard. 2010. Interview with author. November 24.

20 Gutierrez, Jay M., and Alex S. Polonsky. 2007. *Fundamentals of Nuclear Regulation in the United States*. Washington, DC: Morgan, Lewis & Bockius, LLP.

21 Hovey, Robert. 2011. Interview with author. August 29.

22 Ibid.

23 U.S. Nuclear Regulatory Commission. 2010. *Information Digest 2010–2011*. NUREG-1350, vol. 22. Washington, DC: USNRC. August 2–4.

24 Walker, J. Samuel, and Thomas R. Wellock. 2010. *A Short History of Nuclear Regulation, 1946–2009*. Washington, DC: U.S. Nuclear Regulatory Commission. October. http://www.nrc.gov/reading-rm/doc-collections/nuregs/brochures/br0175/br0175.pdf.

25 Gutierrez, Jay M., and Alex S. Polonsky. 2007.

26 Walker, J. Samuel, and Thomas R. Wellock. 2010.

27 Ibid.

28 *The New York Times*. 2011. "Nuclear Agency Beset by Lapses." May 8.

29 *Nuclear News*. 2011. "NRC Chairman Jaczko: A Regulator's Perspective on the Challenges of Today and Tomorrow." July.

30 Walker, J. Samuel, and Thomas R. Wellock. 2010.

31 Lynch, James. 2011. Interview with author. May 25.

32 *Nuclear News*. 2011. "Proposals for a Post-Fukushima Nuclear Safety Framework." July, 17.

33 *Nuclear News*. 2011. "Utility Working Conference: More Space, More Session Tracks at New Venue." October, 88–94.

34 Levis, William. 2011. Interview with author. January 5.

35 PricewaterhouseCoopers LLP. 2011. *Managing Contractors Involved in High Impact Activities*. November. http://www.pwc.co.uk/en_UK/uk/assets/pdf/managing-contractors-in-high-impact-risk-executive-summary.

36 Dlouhy, Jennifer A. 2011. "BP CEO Says Company, Industry Learned from Spill." *Houston Chronicle*, March 8. http://www.chron.com/business/energy/article/BP-CEO-says-company-industry-learned-from-spill-1686063.php.

37 Ellis, James O. 2010.

38 Ibid.

39 National Commission on the BP Deepwater Horizon Oil Spill and Offshore Drilling. 2011.

40 *The Times-Picayune*. 2011. "Former Shell Scientist Named to Head Center for Offshore Safety." March 7.

41 Rogers, Mary J. 2011. *Mission Critical Oil and Gas Operational Reliability and Safety: Learning from the Transformation of Commercial Nuclear Power*. IHS CERA Conference Private Breakfast, March 8, Houston, Texas.

42 Nuclear Energy Institute. 2010. "BP Spill Commission Mulls INPO-Style Group for Oil Industry." *News and Events*, August 26.

CHAPTER 3

Nuclear Safety Culture

Nuclear safety culture has been an important underpinning of the U.S. nuclear energy industry's sustained elevation in safety and reliability performance. Recent industry efforts to define, assess, and improve safety culture have been successful in getting the concept, and the behaviors that promote it, part of everyday life at nuclear power plants across the country. When touring a nuclear power facility, for instance, visitors are inundated with evidence of the nuclear safety culture, from the signage driving in to the plant to the posters on the walls, the safety briefing from security officers, and the start of meetings with safety messages tied to real plant situations. However, the work of nuclear safety culture extends far beyond these visible products to make U.S. nuclear power one of the safest industrial environments in the world.[1]

Although the formal idea of a nuclear safety culture emerged fairly recently in nuclear industry history, many nuclear stations had begun to develop aspects of a robust nuclear safety culture long before the concept was officially defined and examined by regulatory bodies. James Lynch is the vice president of assistance at INPO and has 24 years of experience at the institute. "In the years before we talked about nuclear safety culture, we emphasized nuclear professionalism, conservative decision making, as well as the need for the highest levels of safety and reliability," stated Lynch. "What we now refer to as nuclear safety culture, many of the pieces were there. Defining it gave us all a common language and principles to practice."[2]

This chapter reviews the emergence of nuclear safety culture, how it has been defined and developed across the nuclear energy industry, as well as international use of the concept. The practical results of U.S. nuclear safety culture efforts are discussed in terms of improvements in radiological, industrial, and nuclear (reactor) safety, and advances in nuclear site security and emergency preparedness. Finally, this chapter provides suggestions on how to improve organizational safety culture both inside and outside of the nuclear power environment. Before outlining current nuclear safety culture definitions and practices, however, it is important to review historical events that had a major impact on the nuclear power industry's embrace of nuclear safety culture.

Watershed Events

The nuclear power industry's commitment to the understanding, monitoring, and continuous promotion of nuclear safety culture can be traced back to defining industry events. In a speech at the Regulatory Information Conference in 2008, INPO CEO James Ellis stated that "the most significant industry events, internationally and domestically, that have challenged nuclear safety are due to the absence of or significant weaknesses in two or more of the safety culture principles."[3] The NRC reported that analyses of significant, high-visibility events revealed that weaknesses in safety culture were an underlying root cause of the problems or increased the severity of the problem, or both.[4] The event that appears to have rallied the industry to explicitly define, assess, and proactively foster nuclear safety culture was the reactor pressure vessel head degradation discovered in 2002 at the Davis-Besse Nuclear Power Station.

Davis-Besse event

The Davis-Besse Nuclear Power Station is a single-unit, pressurized water reactor (PWR) located in northern Ohio and owned by First Energy Nuclear Operating Company (FENOC). The Davis-Besse unit ran well and was in good standing with both the NRC and INPO; that is, the plant was not under any additional scrutiny at the time the problem was uncovered. During an NRC inspection of the reactor at Davis-Besse in February 2002, a football-sized hole in the steel vessel that contains the reactor was revealed, leaving only a thin layer of stainless steel. While not imminently life threatening, any additional degradation could have led to a loss-of-cooling accident of the reactor.[5]

Investigations were conducted to determine why the company and the NRC had failed to identify the severe corrosion before it had progressed. The results of the investigations concluded that FENOC had paid inadequate attention to signs of corrosion and had made erroneous assumptions about the need for inspections of the head. The NRC was criticized for tolerating NRC staffing shortages and focusing on more troubled plants, which resulted in fewer inspection hours at the Davis-Besse site.[6]

Further investigations by INPO revealed that management and organizational weaknesses related to safety culture significantly impacted the station's ability to identify and correct the issues that were deteriorating the vessel head. In response to the Davis-Besse event, INPO issued to its members a significant operating experience report (SOER), which required all plants to take action within 90 days. One of the primary recommendations of the SOER was for each plant to conduct a self-assessment on the organization's safety culture.[7]

The Davis-Besse event was a significant emotional event for the industry because the plant was widely considered a good performer and yet it had allowed such a serious degradation to develop. Company and industry leaders concluded that the decisions and actions that resulted in the unknown damaged vessel head were a function of the poor safety culture of the organization. After Davis-Besse, the industry became self-consciously committed to the development of a culture of nuclear safety and reinvigorated and intensified self-critical analysis and conservative decision making, which appear to have resulted in nuclear safety culture as it is known today. Although nuclear safety culture is not a new phenomenon, and many sites had healthy safety cultures before Davis-Besse, the industry has accepted that the intentional promotion of the behaviors and practices of nuclear safety culture are important to plant safety and overall performance. According to Ronald Detwiler, the general manager of quality at the TVA Nuclear Power Group, "It is widely recognized in the nuclear power industry that we realized how important safety culture is in preventing events as a result of the lessons learned from Davis-Besse."[8]

Predating the Davis-Besse discovery by many years was a categorically more serious and disastrous event, the Chernobyl accident in 1986. Chernobyl is discussed here because of its significance to the nuclear power industry, but also because it reinforced the newly forming self-regulatory safety initiatives of the U.S. plants. The devastating impact of Chernobyl firmly established the critical importance of human behavior and workers' attitudes toward safety worldwide.[9]

Chernobyl nuclear accident

On April 26, 1986, the Unit 4 reactor at the Chernobyl nuclear power plant in the Ukraine (former U.S.S.R.) exploded during a test in which the operators had turned off the plant's safety systems and subsequently lost control of the reactor. The explosion blew massive amounts of radioactivity into the environment.[10] The initial explosion killed two workers, and 28 firefighters and emergency cleanup workers died in the first three months after the accident from acute radiation sickness.[11] A series of grave operator errors, in combination with the design of the unit (including the lack of a containment building to stop the release of radiation), contributed to the extensive contamination of the area.[12]

Because operator error was so profound at Chernobyl, the disaster highlighted the critical importance of behavior, attitudes, and values to nuclear safety. Reactor design was also reevaluated. In the wake of TMI, INPO led the way in driving nuclear professionalism and conservative decision making in the United States. The World Association of Nuclear Operators (WANO) was formed in 1989 in response to the Chernobyl accident, with the expressed desire to learn from the successes of INPO and to use Chernobyl's lessons to increase safety worldwide.[13]

Definitions

Defense-in-depth

Underlying the U.S. nuclear power industry's approach to safety is the defense-in-depth philosophy. *Defense-in-depth* is an approach to designing and operating nuclear power plants in a way that creates multiple independent and redundant barriers in order to prevent accidental radiation release. Nuclear power plants in the United States are designed and licensed under the defense-in-depth model. The objective is to compensate for potential human and mechanical failures such that no single layer is exclusively relied upon. Physical barriers, safety systems, and emergency response plans are part of the defense-in-depth approach.[14]

Organizational culture

To understand nuclear safety culture, it is helpful to first define organizational culture. Edgar Schein, professor emeritus at the Sloan School of Management at the Massachusetts Institute of Technology, is considered a pioneer in the study of organizational development and culture. According to Schein,

> The culture of a group can now be defined as a pattern of shared basic assumptions that was learned by a group as it solved its problems of external adaption and internal integration, that has worked well enough to be considered valid and, therefore, to be taught to new members as the correct way to perceive, think, and feel in relation to those problems.[15]

The organization's culture forms over time when a relatively stable group creates a common way of communicating and develops behavioral norms, shared beliefs, and basic assumptions as the members find successful solutions to problems. As the culture develops, the group also determines goals, rewards, and punishments that fit with the evolving pattern of shared basic assumptions.

The nuclear power industry has used this definition to promote a particular culture, one focused on nuclear safety, at each of its plants and the corporate and industry organizations that support them. The goal has been that over time, communications, behavioral norms, and shared beliefs that reflect a commitment to nuclear safety will foster more consistent safety performance across the industry. To this end, industry groups have further defined nuclear safety culture into principles and attributes and have put processes in place to regularly assess and improve nuclear safety culture.

INPO definition

In a member publication distributed in 2004, INPO defined nuclear safety culture as "an organization's values and behaviors—modeled by its leaders and internalized by its members—that serve to make nuclear safety the overriding priority." Together with industry leaders as part of a safety culture advisory group, INPO developed eight principles, with corresponding attributes, for a strong nuclear safety culture. Below are the eight principles with a subset of the attributes included for illustrative purposes:

1. **Everyone is personally responsible for nuclear safety.** This principle includes the need for clearly defined lines of authority,

roles, and responsibilities for safety from the board of directors to the individual contributor.

2. **Leaders demonstrate commitment to safety.** Managers and leaders need to visibly demonstrate their commitment to safety by observing, coaching, and reinforcing standards, as well as correcting deviations from expectations.
3. **Trust permeates the organization.** Organizational trust requires that people are treated with dignity and respect and there is a free flow of information that is timely and accurate.
4. **Decision making reflects safety first.** Decision making must be systematic and rigorous and support safe, reliable plant operation. Operators have the authority and expectation to place the plant in a safe condition when faced with unexpected or uncertain conditions.
5. **Nuclear technology is recognized as special and unique.** This principle requires that the special characteristics of nuclear technology are taken into account in all decisions and actions. Activities that could affect core reactivity are conducted with particular care and caution.
6. **A questioning attitude is cultivated.** Plant personnel demonstrate a questioning attitude by challenging assumptions, investigating unexpected conditions, and considering potential adverse consequences. They do not proceed in the face of uncertainty.
7. **Organizational learning is embraced.** A continuous learning environment is demonstrated by the effective use of operating experience, training, self-assessment, corrective actions, and benchmarking.
8. **Nuclear safety undergoes constant examination.** Independent oversight is valued and utilized, and periodic safety culture assessment results are used as a basis for improvement.[16]

INPO's nuclear safety culture definition and the guidance document, *Principles for a Strong Nuclear Safety Culture*, have taken hold and are used extensively across the nuclear power industry. Nuclear safety culture is part of the organizational effectiveness portion of the biennial INPO evaluation and assessment process in which plants can potentially receive a "strength" or an "area for improvement" rating that they are required to address (ch. 2), among other ratings.

NRC definition

Recognizing the important role that nuclear safety culture plays in plant safety performance, the NRC has also been working to define and assess it. In 2006, the NRC issued enhancements to the regulatory *Reactor Oversight Process (ROP)* to more fully address safety culture. These changes essentially required licensees to periodically assess and proactively manage safety culture as part of their interface with the regulator.[17] More recently, the NRC engaged in outreach efforts, including a workshop on safety culture with multiple stakeholders, including licensees, INPO, the Nuclear Energy Institute (NEI), and other groups and individuals interested in nuclear power in order to develop common definitions and traits.[18] The NRC subsequently released its safety culture policy statement in 2011, in which the NRC reiterates its expectation that "all individuals and organizations performing regulated activities establish and maintain a positive safety culture." The NRC defined *safety culture* as "the core values and behaviors resulting from a collective commitment by leaders and individuals to emphasize safety over competing goals to ensure protection of people and the environment." The following are policy statement traits of a positive nuclear safety culture, as listed by the NRC:[19]

- **Leadership safety values and actions.** Leaders demonstrate a commitment to safety in their decisions and behaviors.
- **Problem identification and resolution.** Issues potentially affecting safety are promptly identified, fully evaluated, and promptly addressed and corrected commensurate with their significance.
- **Personal accountability.** All individuals take personal responsibility for safety.
- **Work processes.** The process of planning and controlling work activities is implemented so that safety is maintained.
- **Continuous learning.** Opportunities to learn about ways to ensure safety are sought out and implemented.
- **Environment for raising concerns.** A safety-conscious work environment is maintained where personnel feel free to raise safety concerns without fear of retaliation, intimidation, harassment, or discrimination.
- **Effective safety communication.** Communications maintain a focus on safety.

- **Respectful work environment.** Trust and respect permeate the organization.
- **Questioning attitude.** Individuals avoid complacency and continuously challenge existing conditions and activities in order to identify discrepancies that might result in error or inappropriate action.

Safety-conscious work environment (SCWE)

A *safety-conscious work environment (SCWE)* is recognized by the industry, INPO, and the NRC as an essential part of a strong nuclear safety culture. A safety-conscious work environment exists when people feel free to raise safety concerns, to their management and to the NRC, without fear of retaliation. Defining a SCWE predates the formal conceptualization of nuclear safety culture. In 1996, the NRC published a policy statement that employers (including licensees, contractors, and subcontractors) need to maintain a SCWE where employees are encouraged to raise safety concerns and employers ensure that there is no retribution for raising safety concerns. In a 2004 guidance document, the NRC explains that while retaliation was already prohibited by NRC regulations, the agency sought to emphasize employers' responsibility for achieving and maintaining a work environment conducive to the reporting of concerns.[20]

Nuclear Energy Institute Task Force

To address concerns regarding multiple safety culture definitions and requirements, the Nuclear Energy Institute (NEI) formed a task force involving interested parties to develop an integrated approach to nuclear safety culture. The NEI task force created a process by which station leadership can continually monitor nuclear safety culture using the breadth of information available in a way that is consistent with the expectations of both the INPO and the NRC. The NEI guidance document places responsibility on the site leadership team to examine the organization's safety culture by regularly reviewing information from the following sources:

- NRC inspection results
- Nuclear safety culture self-assessment

- Industry evaluations (e.g., INPO evaluations)
- Operating experience
- QA/self-assessment/benchmarking/behavioral observations
- Employee concerns program
- Workforce issues, for example, grievance trends, potential SCWE claims, and hostile work environment claims
- Corrective action program evaluations
- Site performance trends[21]

The guidelines include a graded approach to nuclear safety culture assessment clarifying when to use a self-assessment, independent assessment, or a third-party assessment, and what the products and team makeup would entail. The information sources listed above are discussed as ways to identify early indications of potential problems, develop actions to address the issues, and evaluate the effectiveness of the actions. The site leadership team (ch. 11) is recognized as a critical force in monitoring and shaping the station's nuclear safety culture such that it is ultimately responsible for the site nuclear safety culture process.[22]

International Atomic Energy Agency

The International Atomic Energy Agency (IAEA) works to establish international nuclear energy safety norms and recognizes the importance of individual and organizational behavior in nuclear power safety. The IAEA approved the publication of safety standards in 1993, in which principle 3, leadership and management for safety, includes a definition and discussion of safety culture. According to the IAEA, safety culture includes the following:

- "Individual and collective commitment to safety on the part of the leadership, the management and personnel at all levels
- Accountability of organizations and of individuals at all levels for safety
- Measures to encourage a questioning and learning attitude and discourage complacency with regard to safety"[23]

Of note is a recent international effort to enhance previous safety culture definitions to make a more serviceable construct on a global scale. Nils Diaz is the chair of the task force and a former U.S. NRC chairman. The American Society of Mechanical Engineers (ASME) Presidential Task Force on Response to Japan Nuclear Power Plant Events announced that it is working to structure a new nuclear safety construct that would provide a platform for enhancing nuclear power safety worldwide.[24]

U.S. Safety Performance

Safety performance improvements in the United States over the past years have been made in the context of increasing attention to the individual and organizational behaviors of nuclear safety culture. The results are seen in specific improvement areas outlined below.

Radiological safety

The Fukushima events in Japan have raised public awareness and concern regarding radiation exposure. People following the news coverage of Fukushima learned that *background radiation* is the level of radiation naturally present in the environment primarily emitted from earth and the sun. Everyone is exposed to background radiation on a daily basis, but levels vary depending on location; radiation is greater at higher elevations. Individual Americans receive an average of 0.3 *rem* (or 300 millirem) per year from natural sources according to the Environmental Protection Agency.[25] (Europe and Asia measure radiation in *sieverts*; 1 sievert = 100 rem.)[26]

Background radiation levels are not harmful, but there is some debate as to the potential for harm at low levels above background radiation. There is widespread agreement, however, on the dangers posed at higher levels. Based on scientific research and to provide a clear margin for safety, the NRC established conservative maximum exposure levels for the public and for nuclear workers, both of which are tracked continuously at nuclear sites and surrounding areas. Maximum exposure limits are 5 rem per year for nuclear workers and 0.1 rem per year for individuals in the general public.[27]

There has not been a release of radiation from a U.S. nuclear power plant into the environment since TMI in 1979, which was a very low level release. After extensive investigation, no harmful effects have been linked to TMI.[28] Much work has been done across the nuclear industry to greatly reduce the potential exposure to workers inside the plant as well as to the public. *Source term* reduction and control programs reduce and contain the radiation of components in the plant. Barriers and shielding contain the radiation and keep it away from workers. Additionally, a large part of reduced radiation exposure in the plant has to do with worker behaviors and radiation protection programs. These efforts have greatly improved planning and also have increased workers' awareness and sense of personal responsibility for any potential exposure. The industry has shared best practices that reduce exposure for every job. *ALARA (as low as reasonably achievable)* programs are in place at every site, making sure people are educated, trained, and personally accountable for their actions that may put them at any risk. Thus, the industry has trended downward for more than 15 years in worker *collective radiation exposure (CRE).*[29] As a comparison, nuclear workers receiving a measurable dose of radiation received an average of 0.1 rem in 2009, which is less than natural background radiation of 0.3 rem.[30] Figures 3–1 and 3–2 show U.S. plants' average collective radiation exposure per unit for PWRs and BWRs.

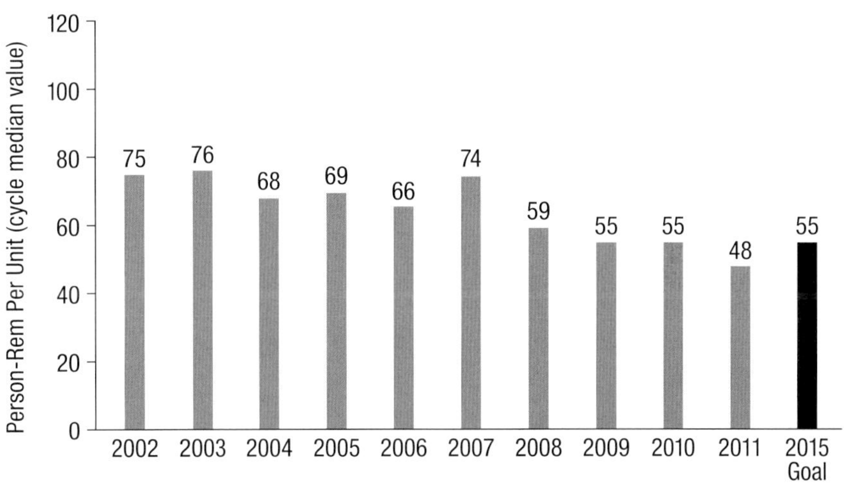

Fig. 3–1. U.S. annual industry average collective radiation exposure—PWR. (Source: NEI; data from WANO.)

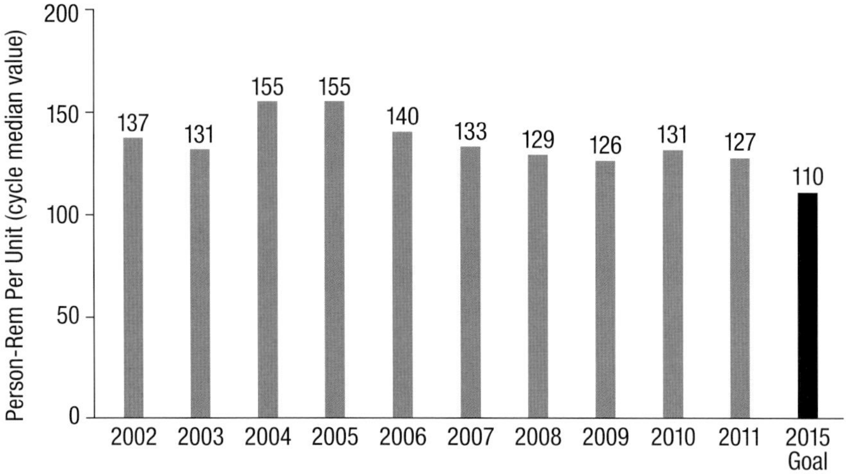

Fig. 3–2. U.S. annual industry average collective radiation exposure—BWR. (Source: NEI; data from WANO.)

Industrial safety

Industrial safety is perhaps the area in which nuclear plants receive the most external benchmarking from other industries. Industrial safety is sometimes referred to as *personal safety* or *personnel safety*. U.S. nuclear plants have become the safest industrial environments in the world.[31] Nuclear power plants track first aids, accidents, illness, injuries, and fatalities—OSHA recordables—like all other industrial, business, and governmental institutions. As shown in figure 1–3 (ch. 1), accidents per 200,000 worker hours were below 0.1 in 2010. It is safer to work in a nuclear power plant than in an educational setting, for example, when comparing occupational injury and illness rates.[32] Nuclear industrial safety accident rates have continued to trend downward even after 2008, when the industry started including all results for supplemental or contracted personnel, as well as employees, in their statistics.

One of the reasons that the nuclear industry is able to maintain such safe work environments is the ownership nuclear sites take for the work and workers on their property. Ownership includes site leaders being held accountable for the safety performance of everyone doing work at their site (ch. 7). Such ownership of safety performance by everyone doing work in the plant, whether company employee or contractor, is required by the NRC as well. Another critical driver of safety is the industry's rigorous application of human performance error prevention tools. *Human performance tools* are simple practices, such as self-check, independent verification, and three-way communication, that, when conducted consistently, greatly reduce the chance that people will make an error. The best performing sites use these tools without fail such that the tools become part of standard operating practices and are written into the actual procedures. Site managers and supervisors monitor workers and one another to make sure people use the tools consistently and respond appropriately with feedback regarding use of the human performance tools.

Nuclear reactor safety

Nuclear reactor safety refers specifically to the programs, practices, and actions potentially impacting reactivity in the reactor core. The nuclear industry has sometimes used the phrase "nuclear safety" to refer to all aspects of safety at the plant, including radiological, industrial, and reactor. In effect, reactor safety refers to how station personnel make decisions and take actions that impact the integrity of nuclear fuel and the reactivity inside the reactor. Reactivity management programs, chemistry effectiveness, configuration control, and control rod positioning are some of the areas that potentially affect the reactor and thus are closely monitored 24 hours a day. Reactor safety is important not only to protect the reactor and fuel, but it ultimately reflects the level of control that the operators and station management have over the unit and the work at the plant affecting the unit. Hence, any unplanned deviation in expected reactor power is duly noted and investigated.

Plants today also have reactivity management programs, which are guidelines and procedures in operations that minimize issues and events that cause unplanned changes to reactivity in the core and maximize consistency and predictability. Industry groups such as the Reactivity Control Review Committee (RCRC) have played a large role in improving reactivity management as they regularly communicate to their site and corporate members about best practices and make detailed and in-depth recommendations (ch. 2).

Nuclear plant owners have also partnered with fuel providers to address fuel cracking trends that were often inherent in the fuel received from the vendor. INPO and EPRI (Electric Power Research Institute) facilitated a partnership with vendors that resulted in greatly reduced fuel failures, which led to reduced reactivity management problems overall.[33] In 2010, 92% of U.S. reactors had zero fuel failures, which was the best performance to date. The current industry goal is zero fuel failures for all 104 reactors.[34]

Security and emergency preparedness

The U.S. nuclear fleet is the safest, most secure, and most prepared that it has ever been for natural disaster or attempted attack. Commenting on the security requirements made prior to September 11, 2001, NRC Chairman Richard Meserve stated, "I am aware of no other industry that has had to satisfy the tough security requirements that the NRC has had in place for a quarter of a century." In response to the attacks on 9-11, the NRC added a series of regulatory requirements for nuclear sites pertaining to security forces, physical barriers, access controls, and coordination with military and law enforcement. For security reasons, many of the changes were not made public. Over the past 10 years, the NRC has made additional post–9-11 requirements related to the capabilities of safety systems, containment structures, and spent fuel pools.[35] Since the Fukushima events, these improvements have been reviewed and U.S. plant capabilities evaluated. The NRC is currently considering the near-term task force recommendations to further strengthen particular areas of emergency preparedness, mitigation, and design-basis seismic and flooding protection.[36]

It is worth noting that five months after the Fukushima disaster, the U.S industry faced a seismic design-basis challenge when a 5.8 magnitude earthquake struck in Virginia near Dominion Generation's two units at the North Anna Power Station on August 23, 2011. Both units tripped automatically in response to a loss of off-site power and were shut down. As designed, backup diesel generators provided power until off-site power was restored.[37] Data recordings showed that the quake had exceeded the plant's design basis for ground acceleration briefly at some locations—a nuclear industry first for an operating plant. In a presentation to the NRC, Dominion Generation reported that none of the reactors' safety-related equipment had been damaged, and nonsafety-related equipment damage was minor and easily repaired. North Anna had also fared well with aftershocks and a hurricane shortly thereafter.[38]

Building Safety Culture

Aspects of nuclear safety culture can seem foreign to those trying to improve safety culture in other industries. When one looks at what managers actually have done at the sites to improve safety culture, however, the work appears comparable and achievable. A simple example is how the INPO's *Principles for a Strong Nuclear Safety Culture* document has been promoted at the sites. The document has been shrunk to a pocket-size booklet, and people carry it with them and use it in discussions. Posters were placed on the walls defining and explaining each principle in detail. Managers start most meetings with a safety discussion tied to one of the principles. Of course, leadership and management have to do much more than talk about and publicize safety principles. They must be role models who demonstrate their commitment to safety through their actions and decisions.

In a speech on ensuring excellence in a global nuclear power industry, INPO CEO James Ellis discussed how to build and sustain a strong focus on nuclear safety culture. Although his recommendations were directed to the nuclear power industry, they are readily applicable to any industry leader:

- **Constantly reinforce safety culture.** Engage people in discussions of the values, standards, and principles that contribute to it. Reinforce successes to guard against complacency.
- **Frequently assess or measure the safety culture in your organization.** Include observations from others you trust who are outside the organization's culture.
- **Make sure the organization clearly understands what constitutes a strong safety culture and its role in promoting it.**
- **Recognize that safety culture is not an all-or-nothing concept.** Every organization has strengths and weaknesses in this arena. Safety culture is a constantly moving continuum in need of daily leadership reinforcement.[39]

At the 2010 Cambridge Energy Research Associates (CERA) oil and gas executive conference in Houston, there was interest in learning from the nuclear industry's safety performance and strong nuclear safety culture. Table 3–1 shows the steps that were presented to improve safety culture in the organization.[40] The final step, fostering self-critical attitudes and a learning orientation, is absolutely essential to building a strong organizational safety culture. Self-criticality is fundamental to identifying weaknesses and making improvements in safety. Such self-critical attitudes need to be apparent from the top of the organization down through the ranks, and include a willingness to admit mistakes and learn from them.

Table 3–1. Key steps to building a strong organizational safety culture

Key Step	Actions to Achieve Results
Leadership commitment to high standards for safety	Gain alignment among the leadership team on the business case for high safety standards; leadership team members then need to commit to specific ways in which they will demonstrate high safety standards and hold one another accountable.
Responding to precursors, small events	Communicate to the organization that you are lowering the threshold on what level of event garners attention and investigation. Consistently respond to low-level events and explain that they are precursors to more serious events and need to be seen as learning opportunities.
Rigorous use of human error prevention tools	Adopt a small set of human error prevention tools, train on their usage, and communicate broadly and repeatedly about the use of the tools. Respond appropriately when the organization uses them consistently and when there is failure to use the tools.
Measuring safety performance using metrics	Regularly measure and track safety performance at multiple levels, including human performance error rates. Communicate results monthly, reinforce progress, and consistently respond to gaps.
Getting workforce input on safety improvements	Safety teams where employees participate need to be well run and credible. Actively seek input and communicate changes made in response to employees' concerns.
Regularly assessing safety behaviors and progress	Use legitimate safety culture assessment processes annually to determine trends and progress made, as well as to proactively identify organizational issues that contribute to safety performance.
Building self-criticality and a learning orientation	Leaders need to foster self-critical behavior by modeling it and by building processes that continually support it. Communicate to the organization the need to be self-critical learners in order to improve performance and grow the organization.

References

1 Institute of Nuclear Power Operations. 2009. Nuclear Safety: Setting the Global Standard. Annual Report. Atlanta, GA: INPO.

2 Lynch, James. 2011. Interview with author. May 25.

3 Ellis, James O. 2008. "Ensuring Excellence in a Global Nuclear Power Industry." Regulatory Information Conference, North Bethesda, MD, March 12. http://www.nrc.gov/public-involve/conference-symposia/ric/past/2008/slides/ellis031208.pdf.

4 U.S. Nuclear Regulatory Commission. 2011. "Safety Culture Policy Statement." NUREG/BR-0500. June. http://pbadupws.nrc.gov/docs/ML1116/ML11165A021.pdf.

5 Walker, J. Samuel, and Thomas R. Wellock. 2010. A Short History of Nuclear Regulation, 1946–2009. Washington, DC: U.S. Nuclear Regulatory Commission. October. http://www.nrc.gov/reading-rm/doc-collections/nuregs/brochures/br0175/br0175.pdf.

6 Ibid.

7 Institute of Nuclear Power Operations. 2006. Significant Operating Experience Report 02-4, Rev.1. Atlanta, GA: INPO. January 27.

8 Detwiler, Ronald. 2012. Interview with author. March 28.

9 Rees, Joseph V. 1994. *Hostages of Each Other*. Chicago: University of Chicago Press.

10 Walker, J. Samuel, and Thomas R. Wellock. 2010.

11 International Atomic Energy Agency. 2012. "Frequently Asked Chernobyl Questions." Newscenter Features, March 21.

12 Walker, J. Samuel, and Thomas R. Wellock. 2010.

13 Rees, Joseph V. 1994.

14 Nuclear Energy Institute. 2012. Safety: The Nuclear Energy Industry's Highest Priority. February. http://www.nei.org/resourcesandstats/documentlibrary/safetyandsecurity/factsheet/safetyhighestpriority/.

15 Schein, Edward. 2004. *Organizational Culture and Leadership*. San Francisco: Jossey-Bass. 17.

16 International Nuclear Power Organization. 2004. Principles for a Strong Nuclear Safety Culture. Atlanta, GA: INPO. 5.

17 Gutierrez, Jay M., and Alex S. Polonsky. 2007. Fundamentals of Nuclear Regulation in the United States. Washington, DC: Morgan, Lewis & Bockius, LLP.

18 U. S. Nuclear Regulatory Commission. 2010. "Revised Draft Safety Culture Policy Statement: Request for Comments." Federal Register, vol. 75, no. 180, September 17. http://69.175.53.6/register/2010/sep/17/2010-23249.pdf.

19 U.S. Nuclear Regulatory Commission. 2011.

20 U.S. Nuclear Regulatory Commission. 2004. NRC Regulatory Issue Summary 2004–x. Guidance for Establishing and Maintaining a Safety Conscious Work Environment. Washington, DC: U.S. NRC. http://www.nrc.gov/about-nrc/regulatory/allegations/scwe-guide.html.

21 Nuclear Energy Institute. 2010. Fostering a Strong Nuclear Safety Culture. NEI 09-07, November.

22 Ibid.

23 International Atomic Energy Agency. 2006. Fundamental Safety Principles. Safety Fundamentals. No. SF-1. Austria: IAEA. http://www-pub.iaea.org/MTCD/publications/PDF/Pub1273_web.pdf.

24 American Society of Mechanical Engineers. 2012. "ASME Task Force Promotes Global Nuclear Safety." Press release. March 15. http://www.asme.org/about-asme/press-releases/asme-task-force-promotes-global-nuclear-safety.

25 *New York Times*. 2011. "Radiation Is Everywhere, but How to Rate Harm?" April 5.

26 Strahlenschutz Dosisangaben ionisierender Strahlung radiation Sievert. 2012. Wissenswertes zum Strahlenschutz mit ionisierender Strahlung. April 9.

27 Walker, J. Samuel, and Thomas R. Wellock. 2010.

28 Ibid.

29 U.S. Nuclear Regulatory Commission. Performance and Accountability Report: Radiation Exposure, Fiscal Year 2010. NUREG-1542, vol. 16, 28. http:/www.nrc.gov/reading-rm/doc-collections/nuregs/staff/sr1542/v16/sr1542v16.pdf.

30 U.S. Nuclear Regulatory Commission. 2011. Information Digest, 2011–2012. NUREG-1350, vol. 23, August, 35. http://www.nrc.gov/reading-rm/doc-collections/nuregs/staff/sr1350/v23/sr1350v23-sec-3.pdf.

31 Institute of Nuclear Power Operations. 2009.

32 U.S. Department of Labor. Bureau of Labor Statistics. 2010. "Incidence Rates of Nonfatal Occupational Injuries and Illnesses by Industry and Case Types." http://www.bls.gov/iff/oshwc/osh/os/ostb2813.pdf.

33 Michal, Rick. 2010. "Kurt Edsinger: EPRI and the Zero Fuel Failures Program." Nuclear News, December.

34 *Nuclear News*. 2011. "INPO: Mixed Results for 2010." June, 30–31.

35 Walker, J. Samuel, and Thomas R. Wellock. 2010.

36 U.S. Nuclear Regulatory Commission. 2011. Recommendations for Enhancing Reactor Safety in the 21st Century: The Near-Term Task Force Review of Insights from the Fukushima Dai-Ichi Accident. Washington, DC: U.S. NRC. July 12, ix. http://pbadupws.nrc.gov/docs/ML1118/ML111861807.pdf.

37 *Nuclear News*. 2011. "North Anna Remained Safe." September, 17.

38 *Nuclear News*. 2011. "Quake Topped Design Basis, but No Vital Damage Found." October, 23.

39 Ellis, James O. 2008.

40 Rogers, Mary J. 2011. Mission Critical Oil and Gas Operational Reliability and Safety: Learning from the Transformation of Commercial Nuclear Power. IHS CERA Conference Private Breakfast, Houston, Texas, March 8.

CHAPTER 4 Operational Focus

Nuclear power industry improvements in safety and reliability accelerated when leaders began increasing the operational focus of their organizations in the 1990s. At that time, a number of state legislatures introduced deregulation in electricity markets and were allowing utility companies to recoup some of their *stranded costs*. While competition breathed life into the utility industry, it put added pressure on nuclear power plants to perform. Some stations were struggling to improve and were on the NRC watch list, receiving additional regulatory oversight and scrutiny. A 1997 study by the engineering and construction firm Washington International Group (now part of URS Corporation) predicted that there would be a premature shutdown of many of the industry's nuclear units because their annual production costs were higher than the prices they would face in the market. Companies were considering decommissioning some of their poor-performing plants well before their operating licenses expired.[1]

Looking back on when he first came to the company, former Exelon Corporation (then Unicom/Commonwealth Edison) Chairman John Rowe stated that he contemplated breaking up the company and selling the nuclear plants. According to Rowe, "We had invested $10 billion and weren't getting a return." Rowe decided to sell the coal plants instead because he would have gotten little for the expensive nuclear plants when they were running so poorly.[2] Having decided to keep the nuclear power assets, it was crucial to get the units operating well and make them more cost-effective.

Christopher Crane, president and chief executive officer of Exelon Corporation, was at the Tennessee Valley Authority (TVA) nuclear generation group in the mid-1990s. "The larger fleets were more poorly operated at that time than the single nuclear sites. Our response to the report from Washington International Group was to focus on operational excellence, increase margins of safety, and drive down costs."[3] In 1998 Crane left TVA for ComEd, where he and Oliver Kingsley, then ComEd's chief nuclear officer, instituted a number of changes that increased operational focus and started the troubled ComEd fleet down a path of performance improvement.

When ComEd merged with PECO (Philadelphia Electric) in 2000 to form Exelon, they created the largest commercial nuclear fleet in the country at a time when large nuclear fleet operators were still unproven. ComEd had already announced that it was closing Zion station, driven in part by the potential major capital expense of replacing its steam generators. While some of the combined PECO, AmerGen (a PECO and British Energy joint partnership), and ComEd plants were seen as cost competitive, a number of them were still vulnerable to being shut down prior to the merger.[4] In the face of diverse and significant challenges, the Exelon Nuclear fleet developed standard operating platforms and started making considerable improvements in safety and reliability to eventually become the best-performing nuclear power fleet in the world.[5] A review of the Exelon Nuclear turnaround (ch. 9) reveals that one of the most significant actions by its leaders was increased attention to *operational focus*.

This chapter examines the concept of operational focus, including how it is defined and demonstrated to help show how operational focus affects organizational and plant performance. A model illustrating a proposed causal pathway from operational focus principles to performance results is presented. The chapter includes a discussion of how to build operational focus in nuclear power plants as well as an example of applying operational focus principles in a fossil generation business and an energy trading organization.

Definition

INPO has found that when plants show significant declines in performance, a key factor has been an unrecognized, long-standing weakness in plant operational focus. Conversely, responding to the early signs of a decline in operational focus is critical to maintaining high levels of plant performance. *Operational focus* is a phrase used by INPO and the industry to refer to the "behaviors and practices of an organization that are necessary to achieve high levels of operational safety and reliability."[6] Rick Purcell, INPO senior VP of industry performance improvement, adds that "operational focus is looking at equipment through the eyes of an operator, who has to operate the equipment under normal and off-normal circumstances. The rest of the organization needs to consider, 'what the operators need to do their job, and what equipment they need under abnormal conditions, such as in an emergency.'"[7] INPO defines operational focus in terms of five principles that underlie the behaviors and practices that make the safe and reliable operation of the unit central to a station's decision making and actions (described below).[8]

Equipment reliability maintained

Leaders, managers, and workers all play a role to achieve high levels of reliability of the equipment needed to monitor and control the plant. This principle includes proactively communicating problems regarding the equipment, making sure problems are resolved in a timely manner, and aligning cross-functional areas to support the resolution of equipment problems.

Low plant risk

When equipment is removed from service or put in a degraded condition, it is done so with a thorough understanding and mitigation of risk to plant operations. Such risks are maintained at a low level by ensuring that work is scheduled and executed expeditiously.

Inadvertent events minimized

The risk of inadvertent events linked to plant activities is maintained at a low level. Personnel need to effectively prevent potential challenges to plant operations and increase focus (and contingencies) when they are planning and conducting complicated, infrequent, or high potential consequence activities.

Expert response to emergent challenges

When emergent challenges to plant operations do occur, operators and other personnel are effectively trained to manage the issues promptly and effectively. In order to do so, roles and responsibilities, procedures, and infrastructure are aligned to support an effective response.

Site-wide roles performed

Both station and corporate personnel need to understand and execute their roles in supporting a strong plant operational focus. From the technicians to the corporate executives, all have an impact on operational focus, from the use of error reduction tools to ensuring adequate resources are provided, for example.[9]

The principles of operational focus have far-reaching implications for the organization as well as for the operations department itself. In practical terms, if the organization is operationally focused, it will attend quickly to operations department concerns and equipment problems that may affect the unit, as well as address any issue that can interfere with the safe and reliable operation of the plant.[10] The operations department has a leading role in reinforcing the operationally focused organization by demonstrating leadership and driving high standards pertaining to nuclear safety, unit reliability, and the performance of plant activities.

Operations Department

Having a high-performing operations department is fundamental to building an operationally focused site organization. At a basic level, site management, operators, and operations leadership need to continuously bear in mind the significance of the role of operations at the station. In the operations department, nuclear operators hold licenses to operate the reactor and, as such, hold the trust of the public, company, site, and the federal government that grants the licenses. The company relies on operations personnel to be leaders in identifying problems with equipment important to plant operation and monitoring, as well as any other concerns, distractions, workarounds, or barriers to operating the unit safely and reliably.

Operators need to serve as the role models for safety and the use of human performance error prevention tools, such as procedure adherence, questioning attitude, and three-way communication, among others. Operators need to feel responsible for decisions that may affect the operation of the unit in any way and ensure that decision making is conducted in a methodical manner that supports safe operations. Operations department management and leadership need to challenge the rest of the organization to address and appropriately prioritize unit safety and reliability related concerns. According to Oliver Kingsley, retired Exelon COO and generation president, "Operations sets the tone for the whole plant. They are the leaders who must demonstrate true professionalism. The control room must be all business. I have always told operators, 'Cut no corners, tolerate no mediocrity, and tolerate no equipment that doesn't work.'"[11]

The operations department also must ensure that the operations training organization is reinforcing operational focus, high standards, and appropriate leadership behavior. The licensed and nonlicensed operator training programs are vital, as they need to provide a continual flow of qualified operators for the unit and department. In order to be successful, operations management works very closely with operations training in all aspects of the training programs, including license class participant selection, classroom and simulator training, and operator requalification. A solid operations department requires a healthy operations pipeline of talent. This benefits not only the department, but also the operators. Some operators will take positions in engineering, maintenance, work management, and other departments at some point in their careers. Operations infiltration into the organization helps ensure alignment around operational focus, but it also significantly deepens the technical capabilities across the organization.

Although the operationally focused organization has a strong operations department, the site needs to maintain a balanced perspective. That is, site leadership needs to balance long-term asset management with operations shift manager priorities. Operations leadership of the site walks a fine line. In the words of Richard Purcell, "Don't overshoot. You don't want the entire site's daily priority to only be the shift manager's daily priority." When site leadership allows operations departments to go overboard, they may become demanding in a vacuum and fail to partner with the rest of the site. As a result, the plant can become too reactive in prioritizing work. The most effective sites have maintenance and engineering management with operations backgrounds who can also provide balance and a technical conscience regarding the licensing basis and design basis for the plant.[12] The site vice president and plant manager play important roles by visibly demonstrating operational focus, while at the same time ensuring that the site organization works together to support safe, reliable plant operation for the short-term and long-term.

Organizational Alignment and Teamwork

Having a strong operations department is a necessary component for building an operationally focused organization, but it is not sufficient. The alignment of the broader site organization and operational teamwork is also essential.[13] In the operationally focused organization, all departments demonstrate sensitivity to the potential impacts to operations and the unit when planning and doing work. Certain plant activities, such as identifying and tagging equipment that is de-energized so that it can be worked on, require a good deal of cross-functional coordination for everyone's safety and for the effective execution of the work. Engineering and maintenance are involved in the monitoring and testing of equipment, which also requires extensive interface with operations. In these and many other situations, personnel from all departments need to maintain an operational focus while they work together effectively to perform all the activities safely and with minimal impact to the unit.[14]

Outside of the operations department, it is easy to get distracted from maintaining operational focus. In a production facility with a variety of important safety concerns, and in a highly regulated environment such as a nuclear power plant, many other things can take center stage. Legitimate potential daily demands on personnel include licensing and regulatory requirements, engineering issues and debates, corporate changes and requests, corrective and preventative maintenance work, schedules and work control problems, security and emergency drills, and personnel and labor relations issues. Of course, many of these demands sometimes do support operational focus.

In the complex mix of potentially competing demands, however, operational focus aligns the organization around a common purpose. Operational focus helps keep in the forefront the nuclear safety culture tenets that the nuclear plant is special and unique and that nuclear safety is everyone's responsibility. Operational focus reminds everyone that they are operating a nuclear reactor and have a control room staffed 24 hours a day monitoring any potential risks to nuclear reactor safety, security, radiological safety, industrial safety, and the production of electricity. Christopher Crane puts a fine point on the alignment of the organization on operational focus. "Whatever we do, we need to keep focused on safe, reliable operations. Everyone needs to tie what they do back to operations and ask what they contribute to operational focus. Ultimately, the operation is the only reason the rest of the organization exists."[15]

Safety and Production Results

Operational focus promotes both safety and production simultaneously by creating an environment where all evolutions of the unit are controlled and the unit performs as expected. To create this environment, personnel have a low threshold for identifying, and appropriately managing, any potential threats to the safe operation of the unit. This also includes operationally focused behaviors by those working across the site, even those who have only an indirect impact on the unit. By having a high degree of rigor regarding potential impacts to reactor functioning, and prioritizing work that improves the monitoring and performance of the reactor, the unit is more likely to achieve its full capacity, barring any problems in the fuel itself. Threats that would potentially take the unit offline are diminished. Hence, an operationally focused organization is a high-performing organization. Figure 4–1 illustrates how practicing INPO's operational focus principles leads to safety and production results via a stable, predictable reactor.

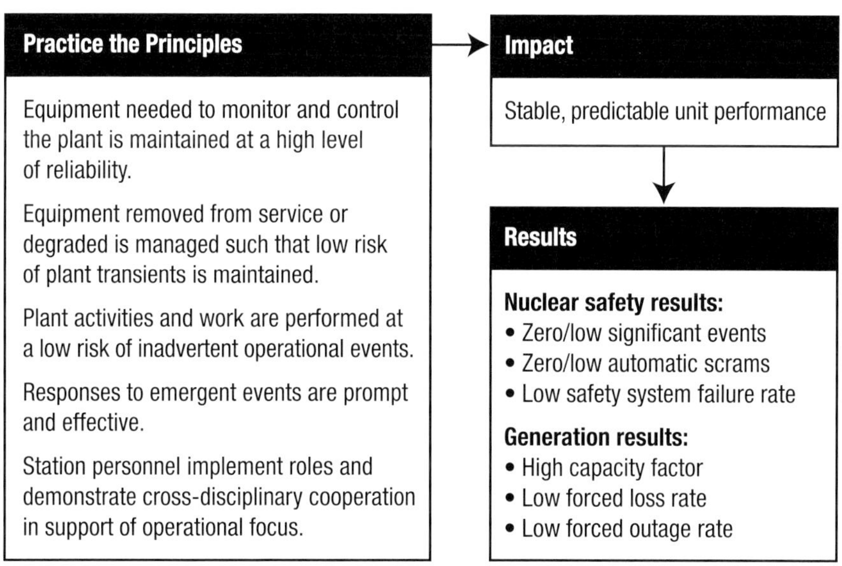

Fig. 4–1. Operational focus pathway to results in safety and production

Building Operational Focus

When Kingsley and Crane came to Exelon, they took many steps that built an operationally focused organization with the help of a number of other leaders and employees across the company. "We decided on standard operating platforms across the 10 sites. Then we standardized procedures, put it all in place, and it had a tremendous positive effect," said Kingsley.[16] Developing standard operating procedures, training personnel, then implementing procedures consistently with rigorous oversight was an important building block for operational focus.

Another step in building operational focus in the organization was to put into place operations-led mandatory standard meetings that focused the attention of the organization on the daily challenges to the operation. The crisp manner in which operations leaders ran these meetings and the standard agenda focusing on safe, reliable operation helped others see that they needed to treat operations as the customer. Daily meetings made it more apparent that plant personnel, such as those in maintenance, engineering, work control, chemistry, and radiation protection, are working to support the safe, reliable operation of the nuclear unit. For example, according to Crane, "Part of engineering's job is to maintain margins of safety in support of reliability for operations."[17]

During these meetings, such as the daily morning report, Crane and the other senior leaders would question the site leadership in such a way that reinforced the critical importance of supporting safe operations. The operations department itself, however, was not off the hook, and was challenged to maintain focus and protect the unit on a daily basis as well. When performance was not meeting expectations, senior leaders would hold frequent follow-up calls or challenge meetings until the site management demonstrated that they were on track.

Building an operational focus in the organization also involves positioning the shift manager as a key part of the management team who leads and drives the resolution of plant operational issues.[18] In multiple cross-disciplinary meetings, the shift manager needs to demonstrate leadership regarding the safe operation of the unit, challenges or potential distractions to operators, and threats to reactor safety. The shift manager is not the only leader, however, and needs to team with other site organizations, take input from others, and listen to their perspectives. Senior site leaders need to develop shift managers who have strong leadership skills but who are not narrow-minded.

Building operational focus includes getting the site leadership team operationally focused and aligned on how operational focus needs to be demonstrated on the team and in the organization (ch. 11). Operational focus can be a powerful alignment tool to build cross-functional teamwork. As members of the SLT work together with a common focus, they can tie their leadership team roles back to how they support the operation of the unit. By looking at issues "through the eyes of the operator," their site organizations will be heavily influenced by their behavior and directions, which builds broader organizational alignment on operational focus.

Fossil generation and energy trading example

Although industries outside of commercial nuclear power can be vastly different from nuclear power in terms of their risk and business models, there are leaders who have applied nuclear operational focus practices to achieve notable performance improvements. William Levis, president and chief operating officer at PSEG Power, successfully implemented a number of modified nuclear processes that resulted in record results for the business.[19]

PSEG Power is a large independent power producer with three main subsidiaries, PSEG Nuclear, PSEG Fossil, and PSEG Energy Resources and Trade. PSEG Power is part of the Public Service Enterprise Group, Inc. (PSEG, formerly Public Service Electric and Gas), based in Newark, New Jersey. William Levis became the president and COO of PSEG Power in 2007 after serving as its chief nuclear officer. Levis has extensive experience in nuclear operations, including his role as the Mid-Atlantic senior VP of operations for Exelon Nuclear. He graduated from the U.S. Naval Academy.

One approach to operational focus outside of nuclear power includes determining the ultimate objective of the operation and focusing on what it takes to achieve and sustain it. In the case of fossil generation, the ultimate objective is the safe and cost-effective production of electricity through the burning of coal, oil, or natural gas. In fossil generation, start-up and shutdown delays are where the business loses money, particularly in a merchant fleet. Across PSEG Fossil, units may be starting up and shutting down 30 times a day. Levis's view is that even though fossil generation has tight margins, the margins can be improved by bettering performance through increased rigor and operational focus. On the other side of PSEG Power, the energy trading organization makes more than 100,000 power trades a year. In the energy trading business, credibility, openness, and transparency are vital to its success. Levis felt that the trading organization could be more effective with increased rigor and a focus on the operation in a way that improves transparency and an open flow of information.[20]

Levis began applying a number of processes to increase operational focus in both the fossil generation and energy trading organizations. His application of commercial nuclear power methods was not welcomed initially, however. According to Levis, "At first, folks fought it like hell. They complained, 'You're going to nuclearize us.'" His response was, "If adding rigor, discipline, and process is nuclearizing, then yes." Levis explained to the management team how the changes were going to help the businesses achieve better results.[21]

Some of the changes included daily operational report outs, regular management review meetings, benchmarking, and formalizing key processes, all with a focus on operational improvement. Levis asked the fossil team such questions as, "How long should it take to start up these units? Who does it faster, and what do they do differently? How long should it take to overhaul these engines?" For the energy trading business, Levis made oversight roles and responsibilities more explicit and independent and information more transparent in order to reduce risk and increase credibility. After implementing these processes and changes, PSEG Power has achieved record generation and reliability results. The fossil plants reduced their startup times significantly.[22]

Levis's experience at PSEG Power is not unique. Outside of nuclear power, there have been various efforts over the years to increase operational effectiveness in general industry. For example, in their bestselling book, *Execution: The Discipline of Getting Things Done*, Larry Bossidy and Ram Charan argue that the leader of an organization needs to be highly engaged in execution, and that it is not a tactical activity that should be delegated. In their model of execution, operating plans and reviews tie the business's strategy to the behavior of the people who implement the strategy. Like nuclear power leaders who build operational focus, the authors recommend regular oversight reviews with candid dialogue, realistic plans, the use of questioning, analysis, and follow-through, and a "relentless pursuit of reality" that is embraced across the organization.[23]

References

1 Washington International Energy Group. 1997. *Nuclear Power Plants and Implications of Early Shutdown for Future Natural Gas Demand.* Washington, DC: INGAA Foundation, Inc. http://www.ingaa.org/File.aspx?id=1358.

2 Rowe, John. 2010. "John Rowe's Bet on Nuclear Energy." *Bloomberg Businessweek*, September 23.

3 Crane, Christopher. 2010. Interview with author. August 6.

4 Washington International Energy Group. 1998. *Need for Natural Gas Increases with More Nuclear Plants Shut Down*. Washington, DC: INGAA Foundation, Inc. 11–12. http://www.ingaa.org/File.aspx?id=540.

5 Kingsley, Oliver. 2011. Interview with author. July 21.

6 Institute of Nuclear Power Operations. 2010. *Principles for a Strong Plant Operational Focus (Preliminary)*. Atlanta: INPO. July.

7 Purcell, Richard. 2010. Interview with author. November 24.

8 Institute of Nuclear Power Operations. 2010.

9 Ibid.

10 Institute of Nuclear Power Operations. 2005. *Performance Objectives and Criteria. Operational Focus*. Atlanta: INPO. May, 37–42.

11 Kingsley, Oliver. 2011. Interview with author. July 21.

12 Purcell, Richard. 2010. Interview with author. November 24.

13 Institute of Nuclear Power Operations. 2005.

14 Institute of Nuclear Power Operations. 2010.

15 Crane, Christopher. 2010. Interview with author. August 6.

16 Kingsley, Oliver. 2011. Interview with author. July 21.

17 Crane, Christopher. 2010. Interview with author. August 6.

18 Institute of Nuclear Power Operations. 2005.

19 Public Service Enterprise Group. 2012. "William Levis: Biography." Corporate Leaders. PSEG Power. April 17. http://www.pseg.com/family/leadership/eog/levis.jsp.

20 Levis, William. 2011. Interview with author. January 5.

21 Ibid.

22 Ibid.

23 Bossidy, Larry, and Ram Charan. 2002. *Execution: The Discipline of Getting Things Done*. New York: Crown Business.

CHAPTER 5 Continuous/Performance Improvement

The nuclear energy industry has taken a number of steps to ensure that continuous improvement is built into U.S. nuclear power plants' processes and culture. The nuclear industry generally refers to continuous improvement efforts as *performance improvement (PI)*. There are several mechanisms at the employee, site, corporate, and industry levels that continually drive improvements across the sites. Underlying these mechanisms is a culture of continuous improvement and learning, which is characterized by a high degree of openness and self-criticality.

The U.S. nuclear power industry's embrace of continuous improvement started with its response to the 1979 accident at Three Mile Island and to the conclusions and recommendations of the investigation. The final report of the presidential commission investigating the accident, the Kemeny Commission, criticized the industry for broader failures that had to be addressed to prevent a similar or more consequential occurrence.[1] Toward that end, one of the major recommendations was the founding of the self-regulatory industry organization, INPO, which U.S. utility executives formed shortly before the commission released the accident investigation report (ch. 2).[2] The act of creating such a self-regulatory body demonstrated that nuclear industry leaders decided to embark on a course of self-improvement.[3]

The TMI accident and Kemeny Commission report are also significant to the emergence of continuous performance improvement programs in nuclear power because they emphasized the need for the industry to pay more attention to the behavior of the people running the plant and the need to prevent systemic breakdowns. The investigation concluded that if the cause of the accident had been only an equipment problem, then TMI would have been a minor incident. It was human error, impacted by organizational and industry-wide failings, which ultimately caused the partial meltdown of the core. In the words of the commission's report, "While the major factor that turned this incident into a serious accident was inappropriate operator action, many factors contributed to the action of the operators, such as deficiencies in their training, lack of clarity in their operating procedures, failure of organizations to learn the proper lessons from previous incidents, and deficiencies in the design of the control room." In addition to these contributing factors, the report noted the lack of a system to ensure that all safety concerns are resolved.[4] As it developed a continuous performance improvement approach, the nuclear power industry focused on these areas initially. More recently, the industry created an integrated model to include multiple programs, then added emphasis on individual and organizational behaviors that support PI.

This chapter presents the industry's early efforts in continuous improvement, highlighting operating experience, corrective action programs, and training. The current performance improvement model and the enhanced focus on behavior are discussed. At the conclusion of the chapter there is a discussion of applying nuclear PI lessons outside the nuclear power industry.

Initial Continuous Improvement Efforts

Initial continuous improvement efforts focused on the conclusions and recommendations of the Kemeny Commission. Operating experience, the corrective action program (CAP), and training were three key areas addressed by the industry, INPO, and the NRC in which formal processes were created. The work in these areas significantly affected the industry's ability to learn from its mistakes and continuously improve its performance.

Operating experience

After TMI, one of the first orders of business for INPO and the NRC was to systematically and continuously gather and distribute operating experience (OPEX or OE) information. The TMI investigation revealed that very similar incidents had been avoided at two other plants in the previous five years, but at the time of the accident, TMI had not learned of the events or how they were effectively handled at the other plants. The NRC did know of these prior incidents, but the information had not been disseminated.[5] The NRC subsequently made significant changes in order to systematically review and share information about the performance of operating plants across the country.[6] It has been suggested that prior to TMI, utilities were operating their nuclear units in many ways like their fossil generation plants, which was in relative isolation from one another.[7]

INPO also established a requirement for all its members (nuclear plant owners and operators) to regularly send to INPO detailed information on all off-normal plant events. Each plant instituted its own OPEX review function to manage these interactions. INPO then screens all off-normal plant events and looks for trends and significance. In 2009, INPO screened 1,900 industry events and classified seven as significant.[8] Subsequent to review and classification, significant operating event reports are then issued with mandatory actions for all plants to address potential vulnerabilities.

Corrective action programs (CAP)

Nuclear power plants are required by law to establish and maintain a corrective action process as part of their overall quality assurance (QA) program.[9] CAP is part of every nuclear plant's license basis and requires that all conditions adverse to quality be promptly identified, documented, reported to appropriate levels of management, and resolved in a timely manner commensurate with the issue. NRC inspectors (and INPO) regularly assess the effectiveness of CAP at every plant.

Corrective action programs have evolved with technology and experience over the years. The industry recognizes the importance of a strong CAP and how crucial it is for all workers to maintain vigilance in identifying, communicating, and resolving deficiencies. In emphasizing the importance of CAP, workers may hear leaders talk about the licensing basis of the plant requiring the prompt identification and management of issues. CAP is also seen as a bellwether for the performance of the rest of the station. In the words of Rick Purcell, INPO's senior VP of industry performance improvement, "Whether you are improving or declining, training and CAP lead the way."[10]

Training

Although nuclear training departments are separate organizations not formally part of the site PI programs, training is seen as a tool to continuously improve performance and was one of the central recommendations from the Kemeny Commission. The nuclear industry has advanced over the past two decades in the development, evaluation, and implementation of training programs. In addition to the accredited and nonaccredited training programs regularly evaluated by INPO (ch. 2), sites have made good use of advances in understanding adult learning, which are applied to continuous training, computer-based training, leadership academies, and other programs. To make sure training is doing its part in site continuous improvement, the following are examples of processes that feed information into training and oversee its effectiveness:

- **Line management ownership and observation of training.** Sites require that their line managers have a minimum number of hours per month during which they observe classroom or other kinds of formal training activities and provide feedback to the training organization. Line managers are expected to demonstrate ownership of training in additional ways, such as ensuring that their employees follow through on training attendance and qualification requirements.
- **Training evaluation requests.** Nuclear plants have a process to evaluate whether current training programs need to be enhanced or new training course(s) should be developed to address particular issues.
- **Training advisory councils.** Cross-disciplinary department, site, and corporate training advisory councils are in place to provide ongoing input to, and oversight of, training programs.
- **Department liaisons.** Personnel in particular departments are designated as the liaisons to training and share responsibility for training outcomes. In the operations department, a full senior position is typically dedicated to operations training.

Current PI Model

Over the past five to ten years, the industry has taken a comprehensive approach to continuous improvement efforts. The multiple programs and processes that were developed over the years have been integrated into a single program of performance improvement. With the help of a working group of leaders and PI managers in the industry, INPO developed *Guidelines for Performance Improvement at Nuclear Power Stations* in 2005. This document provided comprehensive guidance for establishing the overall model for performance improvement, as well as details on the necessary components of a complete and effective program. The industry's model is considered a continuous cycle of problem prevention, detection, and correction (fig. 5–1).[11]

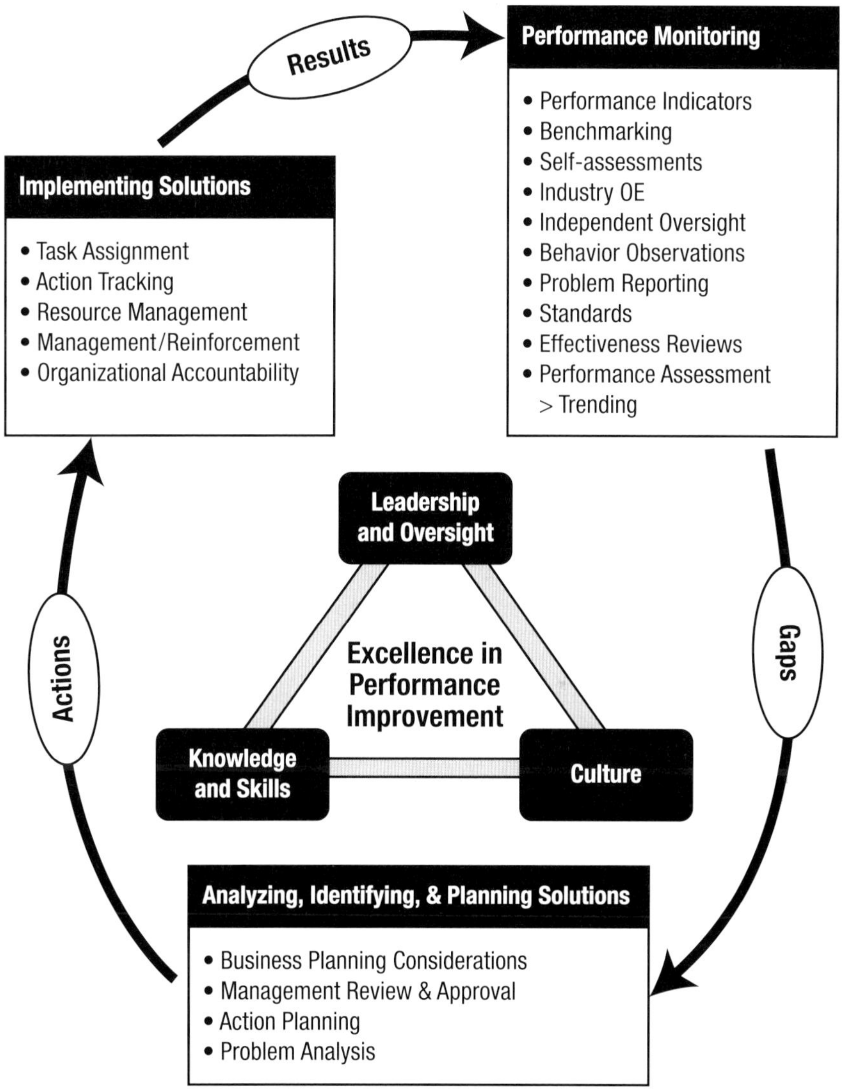

Fig. 5–1. INPO performance improvement model

PI programs and processes

INPO's PI model includes multiple activities supporting performance monitoring, analyzing, identifying, and planning solutions, as well as implementing solutions. The following list is a subset of the elements of the PI model with a brief description of each of the selected elements:

- **Standards.** Standards are part of performance monitoring as they are necessary for identifying gaps. Sites need to avoid comparing their performance only to internal or fleet standards by benchmarking industry standards outside the company.
- **Benchmarking.** Benchmarking is another performance monitoring tool. INPO recommends developing a strategic benchmarking plan, including seeking out best practices in areas of weakness and areas where the site is capable but wants to avoid complacency. Many benchmarking opportunities exist when participating in INPO activities, in addition to traditional formal benchmarking trips.
- **Behavior observations.** Sites use behavior observation programs as a performance monitoring tool. The programs typically have minimum requirements of time spent observing work, and observations are tracked for trends. Adverse trends should be documented in the site's corrective action program. Periodical training on quality observations and coaching helps make the program most useful.
- **Problem reporting.** *Problem reporting* refers to the use of the site's corrective action program (CAP) primarily. It is essential for personnel to have a strong bias for identifying and reporting problems. CAP is a large program with a dedicated manager and team to educate site personnel, track and trend problems, and assist in the analysis and resolution of problems.
- **Problem analysis.** Problem analysis is an element in the analyzing, identifying, and planning solutions part of the model. Sites have formal approaches for analyzing problems in which personnel are trained to be officially qualified to use many of these methods. Root cause, apparent cause analysis, common cause analysis, and event investigation are examples of these tools.
- **Management oversight, involvement, and reinforcement.** This element is part of the implementing solutions section of the model. Essentially self-explanatory, it involves monitoring and tracking

corrective actions, demonstrating a bias for action, ensuring adequate resources, and responding appropriately when personnel are on track with quality work or are offtrack.[12]

In 2009, INPO made a significant enhancement to the PI model by providing detail on the leadership behaviors and organizational characteristics that exemplify the continuously learning and improving organization. This work supported the center triangle of leadership and organizational aspects of excellence in performance improvement. The task force identified behaviors that exemplify treating PI as core business for individual contributors, supervisors, managers, and senior managers.[13] What does it mean to treat PI as core business? According to Richard Cole, industry PI expert with experience at INPO, Exelon, NextEra Energy, and other nuclear companies, it means that the practices of PI become just as important as fixing the physical plant. Cole added, "If human error can take the unit offline, damage equipment, and hurt people (and it can), then PI should be core business and you put in as many defenses as needed to minimize human error. When human errors still occur (and they will), you have adequate layers of defenses to prevent these errors from leading to events."[14] Here is a sample of individual employee-level behaviors that are representative of treating PI as core business:

- Report problem trends to supervisors.
- Report near misses.
- Question unexpected conditions.
- Use knowledge of major station and industry events to internalize work practices that prevent them from recurring.
- Fix minor problems quickly and document them for trending.
- Embrace performance improvement activities as part of daily work habits.[15]

The nuclear PI model has many elements that sustain it. What makes it work, however, is the leadership team's commitment to these programs and the mission behind them. The best sites have strong PI programs and apply the model to their own leaders as well as to the development of their people. They have a learning orientation that they apply to all aspects of running the site, including the people. That is, they are self-critical, seek out and accept feedback, and continually monitor the performance and development of their people (ch. 6). The objectives of current PI programs include ingraining the behaviors associated with effective PI programs into the culture of the organization, where "every individual—leader, manager, supervisor, and worker—embraces a desire for continuous improvement."[16]

Learning from Nuclear Continuous Performance Improvement

The nuclear industry has adopted continuous improvement as a core philosophy in that the leaders and personnel must always be monitoring and improving in order to avoid complacency given the inherent potential for human error and plant degradation. The nuclear industry has learned from aviation, manufacturing, accident theory, and other disciplines, and over the years has incorporated various aspects of continuous improvement into its own models and methods.[17]

The CEO of Entergy Louisiana, Bill Mohl, who has a background in the regulated utility business and unregulated gas and power markets, has an outside perspective on the nuclear continuous performance improvement approach. He sees nuclear power's self-criticality and transparency as differentiating characteristics that are highly useful and is interested in applying nuclear methods to other parts of the utility's business. Mohl attended a monthly management review meeting as an observer to learn more about nuclear PI culture at Waterford 3 Steam Electric Station, located 25 miles west of New Orleans.

According to Mohl, the culture that the nuclear industry establishes enables critical yet constructive discussions of performance. People are expected to be self-critical, acknowledge mistakes, document them in a process, and fix them. The nuclear industry's self-criticality increases the transparency of the operation, which means they are in a better position to manage the risks to the business. "The transparency of the information and the communication of the information are a critical factor to business success. Nuclear works very hard to identify all the risk. Properly communicated, that allows the business owner to take steps to mitigate issues."[18]

In the traditional regulated utility, people sometimes resist increasing levels of transparency—and the accountability that comes with it, presumably because they have not felt the need for the self-criticality and the processes that support continuous PI. Yet Mohl suggests that the time has come for the electricity industry to change its culture. "Customers are more educated and expectations are increasing. The regulators are more educated and holding the utilities more accountable. We need the mentality that we are going to fall behind if we don't continue to get better. And the quicker we can get to be more self-regulating, the better." At the end of the management review meeting at Waterford 3 nuclear station, he told the group how important Waterford 3 is to the New Orleans and greater Louisiana community. Mohl added, "I'm impressed by the humility around the table."[19]

In the process of adopting aspects of nuclear's PI approach, the stumbling block for other industries can be the continual self-criticality. Self-criticality is common in the nuclear power environment and supported by federal law pertaining to the nuclear power safety conscious work environment (ch. 2). That is, people are expected to point out problems and concerns and are protected from any retribution for speaking up.

Bill Levis, COO of PSEG Power, points out that "telling on yourself . . . is not natural." To overcome the apprehension around increasing self-criticality in fossil generation and the trading business, Levis took on a teaching approach. For example, in monthly management review meetings he consistently made managers talk about what did and did not go right, and people learned to be more self-critical and accountable. Levis worked to create an atmosphere of learning that included becoming more self-critical, which was part of an overall effort leading to improvements in generation and reliability (ch. 4).[20]

References

1 President's Commission on the Accident at Three Mile Island. 1979. *Report of the President's Commission on the Accident at Three Mile Island. The Need for Change: The Legacy of TMI.* Washington, DC: President's Commission on the Accident at Three Mile Island. October. http://www.threemileisland.org/downloads/188.pdf.

2 Ellis, James O. 2010. "The Role of the Institute of Nuclear Power Operations in Self-Regulation of the Commercial Nuclear Power Industry." Remarks before the National Commission on the BP Deepwater Horizon Oil Spill and Offshore Drilling. August 25.

3 Rees, Joseph V. 1994. *Hostages of Each Other*. Chicago: University of Chicago Press. 41.

4 President's Commission on the Accident at Three Mile Island. 1979. 10–11.

5 Ibid. 8, 10–11.

6 Walker, J. Samuel, and Thomas R. Wellock. 2010. *A Short History of Nuclear Regulation, 1946–2009*. Washington, DC: U.S. Nuclear Regulatory Commission. October.

7 Rees, Joseph V. 1994.

8 Institute of Nuclear Power Operations. 2009. *Nuclear Safety: Setting the Global Standard* Annual Report. Atlanta, GA: INPO.

9 Gutierrez, Jay M., and Alex S. Polonsky. 2007. *Fundamentals of Nuclear Regulation in the United States*. Washington, DC: Morgan, Lewis & Bockius, LLP.

10 Purcell, Richard. 2010. Interview with author. November 24.

11 Institute of Nuclear Power Operations. 2005. *Guidelines for Performance Improvement at Nuclear Power Stations*. Atlanta: INPO. August. http://pbadupws.nrc.gov/docs/ML1024/ML102460769.pdf.

12 Ibid.

13 Institute of Nuclear Power Operations. 2009. *Achieving Excellence in Performance Improvement: Leader and Individual Behaviors that Exemplify Problem Prevention, Detection, and Correction as a Shared Value and a Core Business Practice*. Atlanta: INPO. September.

14 Cole, Richard. 2011. Interview with author. July 26.

15 Institute of Nuclear Power Operations. 2009. A-1.

16 Ibid. i.

17 Ibid.

18 Mohl, William. 2011. Interview with author. September 12.

19 Ibid.

20 Levis, William. 2011. Interview with author. January 5.

CHAPTER 6 Talent Development and Knowledge Management

The best nuclear power sites apply the same process, rigor, and continuous improvement approaches to talent development and knowledge management as they do other station programs. This dedication has enabled some companies to achieve their strategic objectives by building leadership skills and bench strength, and by putting people with the right leadership skills in the right places. Strong knowledge management programs benefit from new technology and are integrated with recruiting and retention strategies, helping companies keep talented leaders and the precious knowledge they possess.

Oliver Kingsley, the retired COO of Exelon and leader of the Exelon Nuclear turnaround, understood the importance of the people side of the nuclear power business. According to Kingsley, "As a nuclear leader, you have to manage the full equation, not just the operational side or the financial side. The full equation means that 40 percent of your time is spent on managing the operation, 30 percent of your time is spent on the business aspects, and 30 percent of your time should be spent on the people side." At Exelon, Kingsley states they "put in place succession planning processes along with material condition processes."[1]

This chapter is not a comprehensive review of talent and knowledge management (KM) programs, which are widely available inside and outside of the nuclear energy industry. Instead, this chapter discusses selected programs, processes, and lessons learned that have given some nuclear operators a talent and KM advantage in achieving their goals.

Military Recruit Story

Derek Hawes, a 34-year-old operations training supervisor at Sequoyah Nuclear Plant, 16 miles outside of Chattanooga, Tennessee, represents the junior manager and the future leader of commercial nuclear power. He stands six feet two inches tall, bearing a crew-cut and attentive posture. Hawes is ambitious, hard-working, and eager to take on increasing levels of responsibility. Like many other new recruits to nuclear power, Hawes spent time in the military before getting a job at a nuclear power plant. He enlisted and served nine years in the U.S. nuclear navy, where he advanced quickly. He had achieved the level of chief petty officer by working hard and doing the right thing, with a clear picture of what he needed to do to move ahead.

When Hawes decided to leave the navy, he attended job fairs at the air force base in Charleston, South Carolina, where he talked to recruiters from various industries and companies. He accepted a training instructor position at a plant in the southeastern United States. Shortly after his arrival, the plant's training department was placed on probation, and the plant became highly focused on turning around performance issues and earning back accreditation. Although Hawes really wanted to move into a supervisor position, he heard little in the way of potential opportunities for him to advance at the station. He talked with his management about his desire to move up in the organization and was told he may get to a supervisor position in five years. Subsequently, another person was moved into a supervisor position when Hawes had not been given an opportunity to interview for the role.

About a year later, Hawes heard there were opportunities for advancement at TVA, and he started talking with TVA contacts about potential training supervisor positions. He decided to move his family to eastern Tennessee to take a training instructor position at Sequoyah Nuclear Plant. When he informed his managers at his current plant about his decision to leave, they were surprised. His management told him that there were many opportunities for him at the plant, he was on succession planning lists, and that he was viewed as a future leader at the site. None of this talent management information had been previously discussed with Hawes, however. Although he felt bad about leaving the plant, Hawes had made a commitment to join TVA. Within a few months of being at Sequoyah, he was promoted to the licensed operator requalification training supervisor position. He also had an in-depth assessment of his leadership behavior, a feedback conversation that involved his boss and an external leadership development expert, and an individual development plan that spelled out what he needs to do to get to the next level.[2]

Targeted Research on Military Recruits

Properly leveraging the leadership talent of military recruits will become increasingly important for nuclear and other industries as they turn to military veterans for their technical training and leadership experience. A 2010 *Harvard Business Review* article on leadership discusses differences in the way military branches of service prepare leaders for business. The study found that navy and air force executives rely more on processes and are more successful in regulated industries, while executives with army and marine corps backgrounds have an approach that is more decentralized and adaptable. The latter tend to perform better in smaller firms. The authors point out the importance of fit, and encourage companies to understand what characteristics they need from new leaders that they bring into their companies.[3]

When military veterans start work at a company or nuclear plant, their experiences, while becoming integrated into the organization can also affect their success and retention. A recent study of nuclear supervisor effectiveness found that supervisors with military backgrounds may have unique issues related to role transition at commercial nuclear power plants. At nuclear power plants, first-line supervisors with a history of military service rated lower on the following characteristics that are associated with a successful transition to the management team:

- I feel I am treated as a core member of the management team.
- I receive communications from managers above me in a timely manner.
- I am told the reasons behind key decisions and am involved in making decisions.
- I am committed to the organization.[4]

Although a follow-up study is needed to further understand these findings, conversations with military veterans about their experience transitioning to commercial nuclear power provides some ideas to help explain these perceptions about being part of (or not part of) the management team. In the military, it appears that people are accustomed to functioning within a sea of cues showing that they are part of the military "team." In contrast, starting work at some commercial plants can leave recruits feeling as though they have been left to fend for themselves. Military recruits also may have different perceptions regarding their career progression. Military veterans are accustomed to regular, clear-cut, frequent communications and instructions about how and when it is possible to move into a higher rank. As seen in the example with Hawes, without adequate communications about their potential advancement opportunities, it may appear to military veterans that they have little to none.

Talent Development Process and Rigor

Nuclear plants are like many other organizations in their propensity to allow the building blocks of talent development—in-depth prehire assessments for fit and regular career and coaching conversations—to fall by the wayside to technical and daily operating concerns. Hawes's experience is very common; companies lose talented individuals in whom they have invested because their management of talent has been lacking.

At TVA, CNO Preston Swafford appreciates the importance of developing talent and talent management programs to the success of the operation. Swafford and his leadership team have taken steps to jump-start assessment and development and make it an ongoing part of the business. In addition, TVA has defined and communicated the behaviors needed from nuclear supervisors and above. "First you have to clearly delineate the leadership behaviors and skills that you need from talent in order to be successful in an industry that is so unrelenting and unforgiving."[5] Swafford has dedicated substantial resources to leadership development as well as culture change at TVA, including hiring internal leadership and organizational development consultants at each of the sites when other companies are cutting back on talent management staffing. In the late 1990s, Swafford was at Commonwealth Edison, then Exelon, when Oliver Kingsley began the turnaround. Swafford says that he learned a lot from Kingsley and Chris Crane, who partnered with their HR vice president, Rich Landy, to institutionalize the right leadership behavior and put in place the structures of talent management, as well as overhaul other HR areas. "Oliver Kingsley and Chris Crane spent a lot of time on the talent machine, which led to the sustainability of Exelon's performance improvements."[6]

South Texas Project (STP) is a nuclear operating company that has been building up its leadership and employee development programs to support its organizational culture and strategic goals. STP is a two-unit pressurized water reactor (PWR) site located near Bay City, Texas, approximately 90 miles southwest of Houston. STP has some of the newest nuclear plants in the country and is one of the top performers in the industry. Attracting, retaining, and developing talent are critical to STP's overall strategy. Edward Halpin is STP's president and CEO and is an enthusiastic communicator about culture and leadership. STP's leadership and employee development initiatives are part of a broader leadership model in place at the site. Ed Halpin and Mike Meier, vice president of Shared Services, first brought *facilitative leadership*—an approach to organizational effectiveness that emphasizes collaboration and teamwork—to the site in 1994, and since that time they have been creating and nurturing a distinctive culture to help sustain their leadership development efforts.[7] They see their leadership and employee development work (and KM) as an outgrowth of their cultural model. In response to employee feedback and the desire to retain and grow leaders internally, STP created a career development center that provides skills assessments, career coaching, workshops, apprenticeships, and mentoring.[8]

According to Halpin, the work of sustaining a facilitative leadership culture, however, is more process-driven than it appears.[9] There are many structures that support and maintain it, including standard meeting processes, for example. Meeting processes determine what meetings are held and for what purpose, when the meetings occur, who attends, roles and responsibilities, desired outcomes, and the process for conducting the meeting itself. STP employees will even correct visitors when on-site and not following STP meeting guidelines. Establishing processes to support and reinforce desired behavior is what puts the nuclear industry ahead of some others in the talent development and knowledge retention arena.

Exelon Nuclear has demonstrated the positive impact of high expectations, formal process, and rigorous oversight applied to the talent development arena. After the merger of ComEd and PECO in 2000, Exelon Nuclear expanded its management development organization and took multiple steps to institutionalize the talent development process and leadership culture. Leadership assessment, individual development, and succession planning processes were created based on best practices in general industry and the needs of the company. Exelon's talent management processes had regular oversight meetings and metrics to assess results built in to the formal program. Some additional processes included:

- Monthly fleet succession planning reviews focused on functional areas at CNO staff meetings.
- Metrics were gathered and reviewed monthly for trending on succession planning results, turnover of key management, and selection/candidate slate delivery.
- Success maps for key management positions included explicit information on required and preferred background, experiences, and characteristics.
- Biannual fleet succession planning reviews were held covering all key management and executive positions.
- Leadership selection assessments were conducted by external psychologists for all key management and executive candidates.
- External selection assessments/assessment centers were conducted for all first-line supervisor candidates.
- Selection boards were held to review key management selection recommendations.
- Talent development goals were put into executives' and managers' performance reviews in a meaningful way.
- Site senior leadership team assessments and team feedback and development sessions were conducted on a regular basis (ch. 11).

Exelon Nuclear has benefitted from the application of process and oversight to talent development. After putting in place their talent management processes and programs, the company was well positioned in 2004 when they established an operating agreement with PSEG to improve performance at the troubled Salem and Hope Creek plants, located in Hancock's Bridge, New Jersey. In order to effectively conduct the turnaround, Exelon sent 40 of its talented senior managers and succession candidates to Salem and Hope Creek to lead and manage the project at the sites. Exelon's leadership and management of the turnaround at Salem and Hope Creek was viewed as a success, as demonstrated by the large improvements in site generation coupled with an increase in employee morale and safety.[10]

Continuous improvement leadership

One of the tougher nuclear concepts to apply to talent development is the continuous improvement mindset. In order to do so, leaders and managers must be self-critical, open to feedback, and developmentally oriented. It is one thing to be self-critical about a technical matter. It is yet another to be self-critical about leadership and management behavior. Christopher Crane, Exelon president and CEO, applied continuous improvement and rigor to talent development. He stressed the need for regularly using diagnostic tools on individuals and the organizational culture to continually assess potential gaps. Regarding Exelon's talent pool, he stated, "It's not done and it's not perfect and it takes a lot of work."[11]

A demonstration of Crane's continuous improvement approach to Exelon Nuclear's talent development program was seen in a leadership pipeline review project in 2006. This review was undertaken because the senior executive leadership was not completely satisfied with the proportion of "ready now" candidates viewed as capable to succeed in key positions. Given the processes and programs that had been instituted, the senior executives expected stronger output. The reasons for the lackluster succession planning results were not initially apparent. Hence the management development team surveyed and interviewed a cross-section of people throughout the organization, and reviewed metrics and benchmarking information to better understand the gap and potential solutions.

The conclusions were presented to the senior executive team and the site VPs. One of the main findings was that although the leadership assessment and succession planning processes were being followed, the regular career and coaching conversations with potential succession candidates were not happening with enough frequency to have a developmental impact. Moreover, in between formal talent review meetings, little oversight of coaching and development was apparent. "Out of 1,000 questions I get on a typical day, not one is about talent development" captured the perspective of the site VPs interviewed. Senior executives realized that they needed to ensure that they and the site leaders are modeling the right behaviors that the organization is expecting from management. That is, senior leaders need to take time for coaching and developing their direct reports and see that the development programs are being executed at lower levels. Without this demonstration from the top of the organization, the stations will focus more narrowly on technical issues and daily challenges. In addition, Exelon

Nuclear put in place talent development oversight mechanisms that fit into currently scheduled technical oversight discussions at the site.[12]

Developing diverse talent

Commercial nuclear power is not known for its diversity in the workplace or leadership roles. Various efforts have been made over the years to increase the diversity in the pipeline of talent that nuclear draws upon, such as university engineering programs. Table 6–1 shows the gender, race, and ethnicity of nuclear engineering degree recipients for 2009. The Department of Energy, in conjunction with the Office of Economic Impact and Diversity, works with historically black colleges and universities, Hispanic-serving institutions, and Native American colleges and universities to increase minority training in science and engineering. Outreach efforts include conferences, workshops, research, and education awards.[13] Founded in 1977, the American Association of Blacks in Energy (AABE) has a very active scholarship program, drawing applicants from 40 local chapters of the organization across the United States.[14]

Table 6–1. Gender and race/ethnicity of U.S. nuclear engineering degree recipients, 2009. (Source: Fentiman, Audeen. 2011. "NEDHO and Nuclear Engineering Education." *Nuclear News*, January, 45.)

	BS		**MS**		**PhD**	
	Female	Male	Female	Male	Female	Male
Non–U.S. Citizens	2	7	8	23	3	30
U.S. Citizens						
Black/African Americans	2	11	2	2	0	0
American Indians	0	0	0	0	0	0
Asian/Pacific Island Americans	7	15	8	10	2	8
Hispanic Americans	4	8	2	8	0	1
White/Caucasian Americans	49	235	31	137	11	31
Other or Unknown	1	17	0	2	1	0
Totals	**65**	**293**	**51**	**182**	**17**	**70**

Note: Citizenship, gender, and race/ethnicity data were not available for 37 bachelor's degree recipients.

Nuclear site VP perspective

Donna Jacobs, site VP at Entergy's Waterford 3, is taking steps to help grow the nuclear power workforce of the future, which includes a diverse talent pool. Prior to her current role, Jacobs was the senior VP for planning, development, and oversight at Entergy Nuclear. She was previously VP of Nuclear Services at Pacific Gas and Electric's Diablo Canyon plant. Jacobs was also at the Wolf Creek Nuclear Operating Company for much of her career, where she obtained her senior reactor operator certification. She grew up in Kansas and earned her bachelor of science degree in mechanical engineering at Wichita State University and her MBA at Emporia State University. Jacobs's experience of going to engineering school then working in a heavy industrial environment was not a common career path. Her unique path may have started earlier than college, however. Jacobs's father was a car enthusiast and garage mechanic who raised all girls, so she helped him with many of the chores, such as mowing the lawn, changing oil, and fixing cars. Jacobs recalls that her friends thought it was strange when they would call to ask what she was doing, and Jacobs would say, "I'm going to change the oil on the car with my dad." And her friends would ask, "Why?"

Regarding the current and future workforce, Jacobs asks the question, "How do we take a challenging technology of the past and excite people to have the same drive, quality of life, commitment to the community, and passion to protect the health and safety of the public? We have to consider that how we got to where we're at is not necessarily how we'll get the next generation of leaders to where they need to go." At Waterford, they are focusing recruiting efforts on people with ties to the local community. Jacobs is dedicated to leadership development and is looking at high performers deeper into the organization to find people who show signs of leadership potential.[15]

Knowledge Management (KM)

Knowledge management became a major concern in the nuclear power industry as a wave of retirements, especially from the baby boomer generation, was projected to occur at the time when companies were starting the process of applying for licenses to build new nuclear plants. The NRC was also recruiting heavily and staffing up to prepare for an anticipated rise in new build–related construction and licensing applications. Low natural gas prices kept electricity prices low, and the demand for electricity grew at a much slower rate than expected because of the recent recession. With investors more cautious, new nuclear building has slowed. Moreover, some potential retirees are postponing retirement due to losses in retirement funds, and so it appears that the increased demand for nuclear workers may not be as strong as, or occur as soon as, once thought.[16]

The nuclear power industry is not taking any chances, however, and efforts to maintain a skilled nuclear workforce are continuing. For example, 40 community colleges have partnered with nuclear power utilities to implement a certificate program for educating operators and technicians for jobs at nuclear plants.[17] Other partnerships have worked to build up university engineering program participation. In the past five years, enrollment in nuclear engineering programs in the United States has increased substantially. There were approximately 4,800 nuclear engineering graduate and undergraduate students in the United States in 2010, compared to 2,600 in 2005.[18] The International Atomic Energy Agency (IAEA) started a nuclear energy management school specializing in management training for young professionals from countries in the process of developing nuclear power programs for peaceful purposes. The first class was attended by students from 29 developing countries. Of those students, 25% were women.[19]

In addition, the nuclear power industry has moved forward on addressing the impacts of shifting workforce demographics and retirements to address concerns of a loss of critical plant knowledge. Many nuclear companies have focused on knowledge capture by putting into their processes and procedures the hows and whys of what experienced workers know and do.[20] Best practices include the use of technology to capture and distribute information, robust databases, video and multimedia, and coaching and mentoring. At a broader programmatic level, KM programs have these four components:

- Assessment of actual vulnerable areas and individual retirement plans
- Retention of key workers and development of skills
- Recruiting targeted to critical areas
- Knowledge capture and transfer

STP has been proactive and methodical about KM as well as talent development and retention. STP expects to lose 500 people to retirement over the next few years and has been recruiting and staffing, particularly in operations. To help make less-experienced workers effective, STP has increased the content and level of detail in their procedures. But concerns arose over the potentially burdensome, time-consuming impact of thorough and detailed procedures. According to Halpin, "Knowledge transfer takes time. The challenge is, how do you buy time back for people? You need process but you don't want to overburden employees either," says Halpin. STP uses *lean thinking* initiatives drawn from manufacturing to help streamline work and make people more efficient, while in other areas, they may have to enhance processes where knowledge and experience levels have declined.[21]

STP has recently launched "STP wiki" and "STP on demand," two Web-based systems designed to appeal to the newer workers while enhancing knowledge transfer. STP took a senior operations manager and asked him to take over the site's KM efforts. James Mertink is the enthusiastic shepherd for the STP wiki and STP on demand sites. "We want a lot of people to provide input on a subject or a project using current technologies. If we put the information on a wiki-page, it's a time saver and it increases collaboration," according to Mertink. STP also has "ask the specialist" interviews on video linked to the wiki-page. (*Wiki* is Hawaiian for "fast.") STP on demand is a YouTube-like repository of video that includes instructional videos and has many applications for sharing knowledge and, importantly, has the ability to search as well. Some of the instructional videos include how to submit a requisition to purchasing, how to find a vendor diagram, and information on the Terry Turbine. The STP wiki and STP on demand are used for noncontrolled processes that are below the threshold of a procedure or accredited program. The reaction has been very positive. "New people are excited to see the new technology."[22]

The NRC has also been involved in KM and addressing workforce demographics issues by committing to recruiting, staffing, and training in advance of retirements. NRC employees stay with the agency an average of 22 years. The NRC recruited new people to be mentored by the experienced staffers before they retired. The creation of the NRC Knowledge Center provided a Web-based resource that includes social networking with communities of practice and communities of interest, which are used in addition to previous ways of capturing and sharing knowledge (e.g., mentoring, interviewing, writing, and distributing information).[23]

Applying Talent Development and KM Best Practices

The nuclear industry excels in creating processes and procedures to capture and transfer knowledge and by providing rigor and a continuous improvement mindset to talent development. Some nuclear power companies have adopted newer technologies in ways that energize and inform workers as well. The best nuclear sites typically take lessons from general industry leadership development and technology and apply them in the nuclear environment. Hence, the nuclear industry's best practices in these areas are not specific to nuclear power and in most cases had been in use elsewhere.

Many best practices can be obtained from the Corporate Executive Board's Corporate Leadership Council (CLC), which is recognized for its rigorous international research on human resources questions across a broad range of industries. In 2003 the CLC published a major research study on leadership quality and succession management across 276 organizations in 11 different industries. The study concluded that there are seven vital characteristics of the highest performing companies with top-tier leadership organizations.[24] Figure 6–1 presents the seven hallmarks of the top-tier organizations and 34 strategies for achieving results. The 34 strategies are useful in any industry for maximizing talent management efforts. Many have been used in the high-performing nuclear power organizations discussed in this chapter.

Increasing the Chances of Success

By emphasizing the 34 strategies listed here, an organization can significantly increase its probability of being a top-tier leadership organization.

I. Senior Executive Commitment to Development

1. Senior executives are good role models for developing employees
2. Executives believe development is important
3. Line managers make development a priority

II. Organizational Reinforcement of Development

1. Organization modifies development based on organizational strategy
2. Organizational publicly recognizes managers for developing employees
3. Managers held accountable for development in performance reviews
4. Manager bonuses tied to developing employees
5. Manager promotions dependent on developing employees

III. Hiring for Organizational Compatibility

1. Influence of external candidate's future direct reports in hiring decisions
2. Emphasis is placed on external candidate's compatibility with other senion executives in hiring decisions
3. Emphasis is placed on providing information about organizational culture when onboarding new hires
4. Emphasis is placed on external candidate's interpersonal skills in hiring decisions
5. Emphasis is placed on external candidate's compatibility with organization's culture in hiring decisions
6. Reliance on recommendations from senior executives in external recruiting

IV. Exacting Performance Standards

1. Best performers always get promoted
2. Organization "weeds out" worst performers
3. People get ahead based on merit

V. Full Business Exposure for Rising Executives

1. Opportunity to manage large groups of people
2. Opportunity to launch new businesses, initiatives, or programs
3. Opportunity to have significant decision-making authority
4. Opportunity to work on cross-functional teams
5. Opportunity to work in customer/client facing role
6. Opportunity to turn around a struggling business or project

VI. Selecting Successors for Their Leadership Ability

1. Honesty and integrity
2. Ability to recognize and reward achievement
3. Commitment to diversity
4. Ability to identify and articulate long-term vision for future
5. Alignment of skills with future business strategy
6. Ability to allocate resources across competing priorities
7. Ability to clearly communicate expectations
8. Quality of relationships with other senior executives
9. Interpersonal skills

VII. Focus on Scarce Skills and Fit with Position

1. Include people or positions hard to find in external labor market
2. Identify successors for specific positions versus using pools

Fig. 6–1. Corporate Leadership Council's seven hallmarks of top-tier leadership organizations

References

1 Kingsley, Oliver. 2011. Interview with author. July 21.

2 Hawes, Derek. 2012. Interview with author. March 27.

3 Groysberg, Boris, Andrew Hill, and Toby Johnson. 2010. "Which of These People is Your Future CEO? The Different Ways Military Experience Prepares Managers for Leadership." *Harvard Business Review*, November, 80–85. http://hbr.org/2010/11/which-of-these-people-is-your-future-ceo-the-different-ways-military-experience-prepares-managers-for-leadership/ar/1.

4 Fearing, Ben K., Brian D. Juncker, and Mary J. Rogers. 2011. "Selecting and Developing Effective First-Line Supervisors in the Nuclear Power Industry." *Strategic Talent Solutions*, May.

5 Swafford, Preston. 2010. Interview with author. August 4.

6 Ibid.

7 Halpin, Edward. 2011. Interview with author. August 31.

8 Brieden, Laura. 2011. Interview with author. August 31.

9 Halpin, Edward. 2011. Interview with author. August 31.

10 Levis, William. 2011. Interview with author. January 5.

11 Crane, Christopher. 2010. Interview with author. August 6.

12 Exelon Nuclear Management Development. 2006. "Leadership Pipeline Review." Unpublished manuscript.

13 U.S. Department of Energy. 2009. *FY 2009 Strategic Plan for Minority Education Programs*. Washington, DC: Office of Economic Impact and Diversity, Office of Minority Economic Impact. http://energy.gov/sites/prod/files/ED2/minority/documents/Strategic_Plan_2009_-_Minority_Education_Programs.pdf.

14 American Association of Blacks in Energy (AABE). 2012. "AABE Scholarship Eligibility and Criteria." April 24. AABE.org.

15 Jacobs, Donna. 2012. Interview with author. March 26.

16 Fentiman, Audeen. 2011. "NEDHO and Nuclear Engineering Education." *Nuclear News*, January, 45.

17 Nuclear Energy Institute. 2010. "Nuclear Industry's Comprehensive Approach Develops Skilled Workforce for the Future." *NEI Fact Sheet*,

September. http://www.nei.org/resourcesandstats/documentlibrary/newplants/factsheet/nuclearindustrydevelopsworkforce/.

18 Michal, Rick. 2011. "John Gutteridge Moves On." *Nuclear News*, January, 48–49.

19 Seneviratne, Gamini. 2011. "IAEA Launches Nuclear Energy Management School." *Nuclear News*, January, 55–56.

20 Cole, Richard. 2011. Interview with author. July 26.

21 Halpin, Edward. 2011. Interview with author. August 31.

22 Mertink, James. 2011. Interview with author. August 31.

23 Jaczko, Gregory B. 2012. "Strengthening Safety through Managing Knowledge." Prepared remarks for the Honorable Gregory B. Jazcko, chairman, U.S. Nuclear Regulatory Commission, at the South Texas Project Leadership Conference, Bay City, Texas, February 24. http://pbadupws.nrc.gov/docs/ML1206/ML120610044.pdf.

24 Corporate Leadership Council. 2003. *Hallmarks of Leadership Success: Strategies for Improving Leadership Quality and Executive Readiness*. Washington, DC: Corporate Executive Board. http://integratedleadership.com/wp-content/uploads/2011/09/Hallmarks-of-Leadership-Success2.pdf.

CHAPTER 7

Organizational Structure, Accountability, and Outage Management

The nuclear power plant refueling outage has evolved from a 100-day, poorly planned, undisciplined affair, costing companies millions of dollars in overruns, to a 25-day, minutely planned, rigorously executed process saving companies millions in generation and costs. Major changes in outage management have contributed significantly to the U.S. nuclear power industry's record-high capacity factor and lower generation costs. Some sites still struggle, however, to execute refueling outages effectively. Among other factors, the successful nuclear refueling outage is a reflection of the organizational structural characteristics and accountability culture that are seen in well-run sites online. The typical nuclear site organizational structure enables effective operations in an environment where safety and generation demands are high. Accountability culture is critical to effective online operations as well, and when it is lacking, outage performance results suffer exponentially.

This chapter discusses nuclear site organization structure and how particular characteristics of this structure are leveraged to promote effective operations. Accountability will be presented in terms of the internal industry discussion about how to approach accountability without detracting from nuclear safety culture. Lastly, lessons learned in outage management are offered, including the important role of accountability and organizational structure.

Organizational Structure

The typical nuclear site's organizational structure may not be unique or creative, but a combination of factors, consciously applied, has contributed to improved operational effectiveness across the industry. Although structures vary somewhat across nuclear companies, particularly for nuclear fleets versus single sites, there are common patterns and trends associated with high performance. Except in unusual circumstances, the majority of the time the unit is online, and thus the plant organization is focused on safe operation, addressing potential licensing and technical threats to staying online, and conducting corrective and preventative maintenance throughout the plant.

The nuclear power site's organizational structure online supports the notion that the site is a single entity (and some are), and drives the site to function in many ways as one business operation. The structure itself enables each of the following attributes, which are important to the effective running of the plant online and during outages:

- Clear roles with explicit responsibilities
- Clear lines of decision-making authority
- Reporting relationship hierarchy and clarity
- Checks and balances
- Cross-functional teamwork

Explicit responsibilities and clear lines of decision-making authority are dependent upon roles that are clearly defined, simple, and distinct. For example, the maintenance manager is responsible for the safe and effective execution of maintenance work at the station. The maintenance department is typically the largest on site (the operations department is the next largest) and includes mechanical, electrical, instrument controls, union-represented and/or nonunion-represented craft workers, and the supplemental maintenance workforce, including contractors. The superintendents of these departments (mechanical, electrical, etc.) have full responsibility for the maintenance work and employees in their areas. A typical maintenance annual budget is $45 million with 250 employees. As such, the maintenance manager position is a highly demanding role and can be part of a good preparation for the plant manager position.

Nuclear organizational structures also tend to have reporting relationship hierarchy and clarity. Generally speaking, nuclear plants make judicious use of matrix organizational structures. Nuclear site leaders and managers have a high need to manage risk by knowing definitively who is doing what, when, where, why, and how. Nuclear power sites have a low tolerance for risk stemming from diffusion of roles and responsibilities. Hence, the nuclear organizational structure is more hierarchical with direct hard-line reporting relationships and fewer dotted-line reporting relationships. As seen in the maintenance example, the superintendents of mechanical, electrical, instrument controls, and the support/supplemental workforce report directly to the maintenance manager. The maintenance manager and other key leaders who run the plant (e.g., operations manager, work control manager, radiation protection manager, and chemistry manager) report to the plant manager, who in turn reports to the site VP, who is responsible for the entire site (see fig. 7–1).

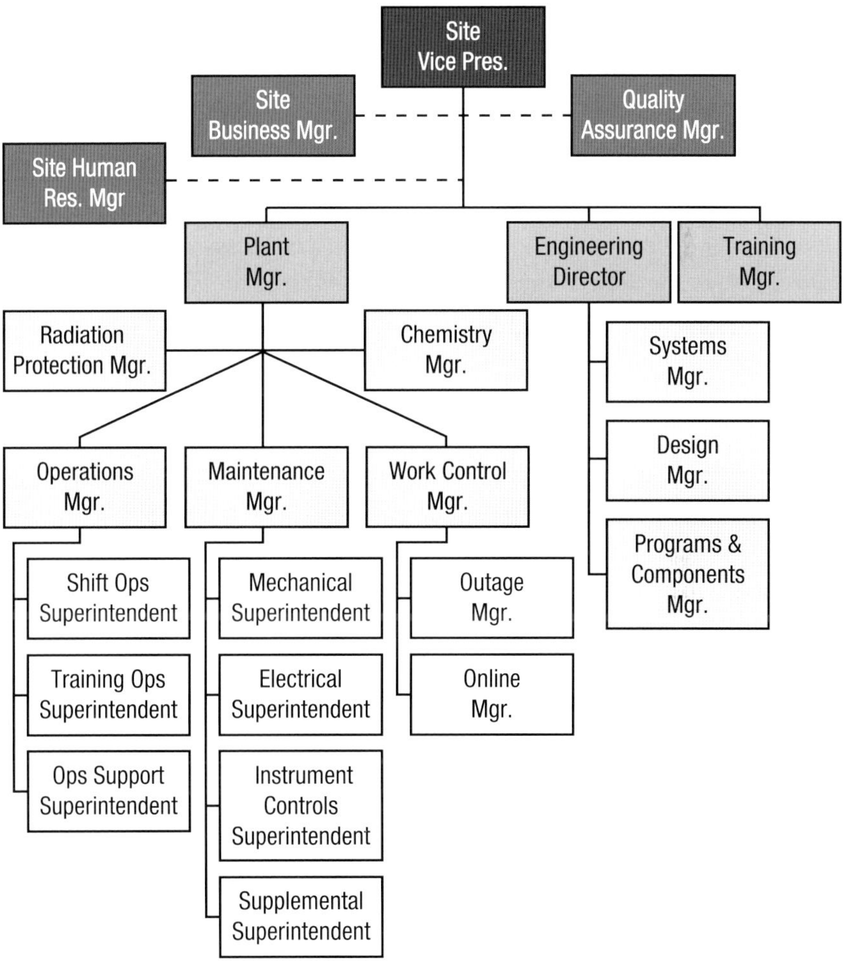

Fig. 7–1. Typical nuclear power site organizational structure—senior managers

Figure 7–1 illustrates the bias toward hard-line reporting relationships up through the site VP, who reports to the senior VP of operations in a fleet, or CNO in a single-site company. The only true dotted-line direct reports to the site VP are the QA manager, site business manager, and site HR manager, where the latter two may also be hard-line reports in smaller fleets or single sites. The site engineering director and training manager typically have dotted-line reporting relationships to corporate functional area leaders.

The structure also supports checks and balances and cross-functional teamwork, both of which are vital to a well-run nuclear power plant. Departments such as operations, maintenance, work control, radiation protection, and chemistry must work together to operate and maintain the plant, and so report to the plant manager. As a counterweight to production and a technical conscience for maintaining the plant design basis, the director of engineering typically reports to the site vice president. Quality assurance, also called *nuclear oversight*, reports to the corporate quality organization and serves in an independent, continual assessment, and oversight capacity (ch. 2). Hence, QA has a dotted line to the site VP, who is the site QA manager's chief customer, while the QA manager's boss is a corporate VP. Engineering and QA provide checks and balances to the production managers.

With training, engineering, QA, human resources, and finance reporting to the site VP, and the plant leaders generally reporting to the plant manager, the site VP and plant manager play a critical role in bringing the broader team together to work cross-functionally to serve the site's needs as a whole. For example, the site VP ensures that engineering works cooperatively and supportively with operations and maintenance managers to maintain the unit on a daily basis and support the long-term asset management of the unit. Cross-functional teamwork is essential to the effective functioning of the site organization. Cross-functional teamwork is necessary for the production managers who report to the plant manager, but it is equally important for the larger group of managers and senior leaders, who have more diverse goals, roles, and responsibilities. The organization's meetings, schedules, and accountabilities also need to support cross-functional coordination, challenge, and teamwork.[1]

Accountability

Historical perspective

The nuclear power industry's emphasis on accountability is an important part of achieving and sustaining high performance in safety and reliability. Over the years, however, the nuclear industry has undergone changes in its approach to accountability in search of a model that is effective yet supports nuclear safety culture. There remains relative variability in how nuclear sites treat accountability. While some sites and companies have adopted a model and execute it consistently, others lack an effective approach. When most nuclear sites first began operation in the 1970s and 1980s, historical evidence suggests that they were run by the utilities much like they ran their fossil units at the time; that is, they were inward looking and production-oriented.[2] In line with the utility industry, if there was accountability, it was likely in the form of discipline in response to flagrant infractions of company policy or legal violations on company property.

The creation of INPO after the accident at Three Mile Island brought more of a nuclear navy influence in many areas across the nuclear power industry and an increased emphasis on standards, management oversight, and accountability. INPO's leadership was primarily comprised of retired navy admirals and other officers. Retired nuclear navy officers also started taking leadership roles at commercial nuclear sites and companies throughout the United States. Such leaders were heavily influenced by Admiral Hyman G. Rickover, referred to as the father of the nuclear navy, and known for his high standards and ruthless intolerance for failure to work toward the standard of excellence.[3] "The navy brought much needed discipline and adherence to process to commercial nuclear power, as well as a 'thou-shall-do-what-I-say' leadership style in some parts of the industry," according to Jeffrey Merrifield, senior VP at the Shaw Group and retired NRC commissioner.[4]

During the wave of site turnarounds starting in the 1990s, the visible increases in individual and group accountability for performance results were keys to rapid improvements at problem plants that were in danger of being shut down by the NRC—or by their owners (ch. 4). When high accountability was initially being emphasized, it tended to be swift and public in order to have an impact on the broader organization. Some industry professionals started expressing concern regarding the potential negative effects of high, harsh, and visible accountability demonstrations on the openness of people in the organization. In 1996, the NRC published a policy statement on the need for nuclear power operators to maintain a safety conscious work environment (SCWE), where people feel free to raise concerns without reprisal.[5] Over time, industry executives have increasingly discouraged the use of harsh public accountability in order to preserve a safety conscious work environment where people are not afraid to admit mistakes and openly discuss problems.

According to INPO, deficiencies in accountability, along with supervisors not reinforcing standards and expectations, remain among the most frequent causes of problem areas in plant evaluations.[6] Seemingly in reaction to the harsh accountability approach of the past, some site leaders have backed away from formalizing an accountability model and from holding people and the organization accountable. In some cases, the pendulum may have swung too far in the other direction. Some site leaders take a passive approach because they do not want to set a tough, punitive tone in the organization. Others argue that they want the employees to hold themselves accountable. With a vague or nonexistent accountability model, some site leaders may unintentionally neglect their ultimate responsibility to ensure that people in their organization actually do what they are supposed to do and that managers and supervisors respond to gaps in behavior and results. At these sites, management gradually becomes tolerant of poor performance and standards decline.

James Reason, who first proposed the concept of *just culture* as an important component to safety culture, expresses concern about the prevalence of "no-blame culture," which would currently be considered a misapplication of his work.[7] International working groups on aviation safety have sought to educate and correctly apply the writings of Reason and others on safety culture and accountability. Just culture is characterized by an atmosphere of trust in which people are encouraged to provide safety-related information and in which the organization has clearly defined acceptable and unacceptable behavior. Reason emphasized the role of systemic organizational factors when errors are made, but did not overlook the actions of individuals. Instead of focusing on the individual alone, however, other factors are examined in order to identify organizational weaknesses and to learn from them.[8] INPO includes a method for examining individual and organizational factors when analyzing an error in its guidelines document on human performance tools.[9]

As the industry shapes its accountability model, leaders and managers need to uphold their responsibilities to the organization to make sure that the behavior of people on-site support the safe and reliable functioning of the plant and those behaviors that are detrimental are sufficiently addressed. Leaders also have a responsibility to ensure that organizational barriers to errors, such as processes or procedures, are of high quality and function effectively, and that responsibilities and unacceptable behavior are clearly defined.

Applications of accountability

Tim Rausch is the CNO at PPL's Susquehanna nuclear station in Salem Township, Pennsylvania. He was previously a site VP at the Oyster Creek nuclear station in New Jersey and fleet engineering director for Exelon Nuclear. Rausch's approach to accountability reflects an appreciation of organizational and safety culture ideas. According to Rausch,

> It is very difficult to change an organization's culture and create a culture of accountability. I'm not an expert, but I know that when you try to attack accountability, you cannot start at the wrong spot. First, you need to clearly establish responsibilities so people know what they will be held accountable for. Then you make sure the processes and programs are effective and educate people on what good looks like. Then you bring in an accountability model that's implemented fairly and with respect for the people. When there is an error, you ask a lot of whys, to assess organizational and programmatic gaps, as well as individual behavioral gaps. In that environment, people are more apt to take ownership and be self-motivated, and will impose on themselves a desire to continuously improve.[10]

There are a number of resources and tools for promoting effective accountability. At the large dual-unit South Texas Project (STP) in Bay City, Texas, accountability is seen as an important part of the organization's success. CEO Ed Halpin and his leadership have used Crucial Conversations techniques to help people have those difficult conversations in an effective manner without damaging safety culture. Crucial Conversations training provides a methodical, yet heartfelt approach to challenging conversations and explains the costs of avoiding such conversations.[11] "With Crucial Conversations tools, we make it safe for people to get the conversations on the table. It provides a structure for the organization, and also improves predictability."[12]

Another resource is in Patrick Lencioni's widely read book, *The Five Dysfunctions of a Team*. Lencioni includes "avoidance of accountability" as one of the five dysfunctions that hold back teams and organizations from achieving results. In this context, accountability refers to the willingness of team members to call their peers on performance or behaviors that might hurt the team or the ability of the team to achieve their goals. Avoidance of accountability is a vital dysfunction that must be overcome for the team to achieve collective results. Lencioni points out some ways to overcome the avoidance of accountability, including publishing goals and standards, having simple and regular group and individual progress reviews, and team rewards.[13]

Accountability is recognized across the nuclear power industry as a crucial part of effective nuclear operations. INPO includes "healthy accountability" as one of the leadership fundamentals to achieve and sustain excellent station performance. Sites that perform better in both safety and generation consistently apply a healthy accountability model. Leadership behaviors that support a healthy accountability are captured in the INPO's description of leadership fundamentals, described in *Leadership Fundamentals to Achieve and Sustain Excellent Station Performance*.[14]

Healthy accountability

- Establish and consistently reinforce that coaching, feedback, and acting to correct performance shortfalls are expectations for workers.
- Openly acknowledge and be accountable for leadership shortfalls, and share the actions taken to correct mistakes.
- Solicit challenges to group thinking by asking questions such as, "What are we missing?" and "Where are our blind spots?"
- Provide ongoing feedback and coaching to reinforce adherence to defined standards and expectations.
- Ensure positive consequences for employee adherence to defined standards and expectations and negative consequences for nonadherence.
- Take prompt and effective action to correct performance shortfalls.
- Create and reinforce the expectation to meet the commitments made to others.

- Assist workers in being accountable through establishment of clear ownership of and due dates for priority improvement actions.
- Ensure that employees at all levels are provided candid and meaningful performance feedback on an ongoing basis.
- Use a variety of methods, including performance indicators, independent feedback, and direct observation in the field, to maintain an integrated picture of performance.
- Observe meetings, visit work locations, and routinely place eyes on the plant to gain personal perspective on issues. Obtain direct feedback from employee issues and concerns, to identify and help eliminate barriers to success.
- Frequently communicate current performance to internal and external stakeholders, and solicit feedback on progress and challenges.

Nuclear application of an accountability model benefits from an organizational structure that reinforces clear roles and responsibilities and direct lines of authority. These characteristics of clear roles and responsibilities, individual lines of responsibility, and accountability are also crucial to executing a safe, timely refueling outage.

Outage Management

In the United States, the bulk of the site organization commits full-time to execute the site's nuclear refueling outage. During the outage, the organization transforms completely while most employees take on an "outage job." The focus during the outage is on one goal: the safe completion of the outage—the refueling, maintenance work, and capital improvements according to the extensive, detailed schedule. Refueling outages typically are conducted every 18 or 24 months per unit, depending on the requirements of the *fuel cycle* of the reactor. Most people put aside their regular jobs for the duration of the outage and commit to a new work schedule and demanding conditions for the duration. Over 1,000 contractors are processed for access to the plant, badged, and managed on-site for the outage. Monstrous and minute pieces of equipment are brought in and/or carried off-site, sometimes by rail. The outage control center (OCC) is staffed 24 hours a day to lead and manage the outage, while operators continue to monitor the unit(s) and run the operations control room. The schedule of activities includes extensive detail, even more so around *critical path* activities, the completion of which have a direct impact on the successful, timely progression of the outage. When done right, a refueling outage is a well-oiled machine as a result of excellent planning, scheduling, and execution.

Nuclear outages have improved greatly over the past 10 years, and sites are now setting records for duration, safety, and cost (fig. 7–2). Even when major equipment is modified or changed out, in addition to the removal of the unit's old fuel and installation of the new fuel, many plants are showing record efficiency. Outage preparation, staffing, scheduling, and operation of the OCC present opportunities to learn techniques that can be applied to other industries that have time-sensitive, large projects with high safety demands.

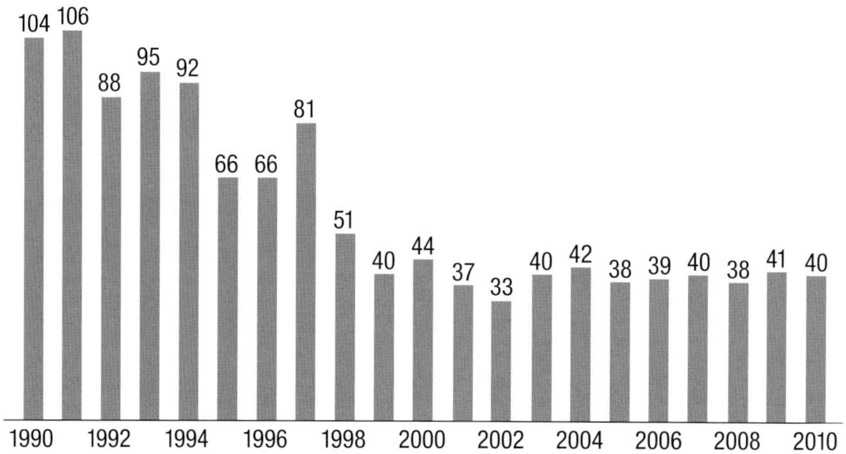

Fig. 7–2. U.S. nuclear refueling outage days (average). (Source: 1990–98 EUCG, 1999–2010 Ventyx Velocity Suite/Nuclear Regulatory Commission.)

Preparation

The importance of refueling outages is evident from the fact that sites spend approximately 25% of their fuel cycle budget on the outage alone. A single-site operating company may have a $300 million budget for two years of operation (24-month fuel cycle) and spend $75 million on their refueling outage.[15] Les Guthrie, retired from Exelon as the VP of outage management, currently consults across the industry on outages and work management and serves on various sites' nuclear safety review boards. While at Exelon, he had multiple operations and maintenance leadership roles, in addition to his outage position, and was deeply involved in numerous record-breaking outages. Exelon is known for having many of the most efficient, best-run, and shortest outages.

Before Guthrie will talk with site leaders about the nuts and bolts of how to conduct successful outages, he says,

> You have to start with the right overall plant management philosophy to be successful at outages. Does senior plant management understand the business they are in—making electricity safely, operating the plant, maintaining the plant, and doing efficient outages? Outages have to be part of core business, not just something you start working on six months before the outage starts. Too many senior management teams get caught up in the issue of the day and fail to pay sufficient attention to outage planning when they are online.[16]

The key to a successful outage is what is accomplished before the breaker opens. Guthrie points out that "95% of the planning and preparation has to occur before the outage starts." He says planning truly starts five years beforehand with long-range planning, long-term asset management (in which it is determined and planned out how and when obsolescence will be addressed), major modifications, major capital improvements, and power uprates. After the long-term plan is complete, planning continues starting with a *baseline schedule*. The baseline schedule is the shortest possible refueling outage given a particular fuel and plant. A typical baseline may be 13 days and 6 hours, with all maintenance, resources, staffing, and required testing built in and *logic tied*. The outage expands from there but with rigorous scrutiny and planning that requires excellent cross-functional teamwork.[17] Preparation also includes making sound decisions about outage leadership roles, defining roles and responsibilities, finalizing work packages, and verifying the level of preparedness with sufficient detail.

Outage control center (OCC)

For the most consistently efficient, safe outages, the OCC is the place where key managers run the outage. Scheduling changes and issues are discussed and processed through the OCC. For the best results, the people who are running the OCC need to have been heavily involved in the development and preparations for the outage. There is a defined hierarchical organization structure for the OCC as well, with clear roles and responsibilities. Prior to the outage, the shift outage managers (SOM) for the day and night shifts (SOM-days, SOM-nights) are the top leaders in the OCC. Together with their OCC teams, they conduct daily schedule reviews and contingency plan reviews.

People working in the outage need to follow the 10-30-60 rule, which means that if the worker cannot resolve an issue in 10 minutes, he or she needs to contact the supervisor. If the supervisor cannot resolve it in 30 minutes, then he or she notifies the OCC. If the OCC cannot get it back on track within the hour, then the OCC needs to notify the plant manager and site VP.[18] Frequent status meetings are held on the outage, sometimes daily and sometimes hourly, depending on how well the outage is progressing.

Using Nuclear Structure, Accountability, and Outage Management Lessons in Other Industries

Nuclear organizational structure principles can be applied to other operations or generation units where clear lines of authority, tied to increased accountability and ownership, would improve effectiveness. Outage management rigor pertaining to preparation, scheduling, and teamwork can be applied to time-sensitive large capital projects. Below are the viewpoints of executives who have both nuclear and fossil operations backgrounds and a perspective of how to apply these nuclear best practices outside of the nuclear power industry.

Preston Swafford is currently the CNO of TVA and has also served as the senior executive for fossil operations and transmission and distribution for Exelon. Swafford has seen that nuclear concepts of accountability have to be applied differently outside of the nuclear power industry. "You need the ability to create a high accountability culture without losing the rank and file, without taking away from a safety culture. You have to use a variety of techniques and it is a little slower." His current approach at TVA includes clearly delineating behavioral expectations and defining the model for healthy accountability. In addition, Swafford uses regular management review meetings as well as leadership assessments to gauge and improve accountability and alignment.[19]

Mark Schiavoni, SVP of fossil, renewable generation, and fuels at Arizona Public Service (APS), has a navy nuclear background. He has also served as plant manager at the LaSalle nuclear station, and as president of fossil generation for Exelon. Schiavoni cautions that fossil generating plant leaders fear that the rigidity of nuclear processes and practices will take away their ability to think and be responsive when doing their job. Moreover, fossil plants do not have the budgets and people to do many of the nuclear processes. Thus, it is necessary to be selective about what is transferred from the nuclear power industry. Schiavoni has applied a more nuclear accountability model to his organization, and he selects things to go after—stretch goals where the organization can strive to be much better. Instead of the nuclear mantra of striving for excellence in everything, they pick something where the people can see the benefits of it, then go after the next stretch goal.[20]

Bill Levis, COO at PSEG Power, has used many aspects of his nuclear experience in other parts of the business, suggesting that the transparency, rigor, and discipline increases credibility, which helps the other businesses in many ways. At PSEG, his team has applied nuclear outage techniques to improve fossil plants' start-up and shutdown times and overhauling engines. Fossil plant start-up times cost the company money. Though operating margins are small in fossil, Levis says nuclear approaches can improve them.[21]

References

1 Rogers, Mary J., and Colleen M. Walker. 2010. "Evaluation of Leadership Teamwork Interventions at DTE Energy—Fermi 2 Nuclear Power Station." *Energy Workforce*, Summer.

2 Rees, Joseph V. 1994. *Hostages of Each Other*. Chicago: University of Chicago Press. 23.

3 Ibid.

4 Merrifield, Jeffrey. 2011. Interview with author. May 10.

5 U.S. Nuclear Regulatory Commission. 2010. "Revised Draft Safety Culture Policy Statement: Request for Comments." *Federal Register*,

vol. 75, no. 180, September 17. http://69.175.53.6/register/2010/sep/17/2010-23249.pdf.

6 Institute of Nuclear Power Operations. 2010. "Sharing Best Practices in Accountability." *Organizational Effectiveness Digest*, March.

7 Reason, James. 2004. "Foreword." In *A Roadmap to a Just Culture: Enhancing the Safety Environment*. GAIN Working Group E. Alexandria, VA: Global Aviation Information Network. September, vi. http://flightsafety.org/files/just_culture.pdf.

8 Reason, James. 1997. *Managing the Risks of Organizational Accidents*. Burlington, VT: Ashgate Publishing Company.

9 INPO. 2007. *Human Performance Tools for Managers and Supervisors*. Atlanta, GA: INPO.

10 Rausch, Tim. 2012. Interview with author. April 26.

11 Patterson, Kerry, Joseph Grenny, Ron McMillan, and Al Switzler. 2002. *Crucial Conversations: Tools for Talking When Stakes Are High*. New York: McGraw-Hill.

12 Halpin, Edward. 2011. Interview with author. August 31.

13 Lencioni, Patrick. 2002. *The Five Dysfunctions of a Team*. San Francisco: Jossey-Bass. 212–215.

14 Institute of Nuclear Power Operation. 2007. *Leadership Fundamentals to Achieve and Sustain Excellent Station Performance*. Atlanta, GA: INPO. September.

15 Guthrie, Lester. 2011. Interview with author. July 21.

16 Ibid.

17 Ibid.

18 Ibid.

19 Swafford, Preston. 2010. Interview with author. August 4.

20 Schiavoni, Mark. 2010. Interview with author. August 19.

21 Levis, William. 2011. Interview with author. January 5.

CHAPTER 8 Nuclear Industry Leadership

Leading the U.S. nuclear power industry's significant improvements in safety and production over the years were a number of highly committed individuals with full appreciation of the potential and the risks of the technology. There were also many dedicated people in nuclear power who worked hard, long days over their careers with little fanfare. There was a lot of work to be done. Despite the solid performance of a few stations, many more performed poorly and some were off-line for numerous months due to technical or regulatory problems, or both. Some plants were ultimately shut down permanently.[1] As can be seen in figure 8–1, the average generating capacity factor was a meager 56% in 1980 but greater than 91% in 2010. Without a significant and sustained improvement in capacity factor, as well as safety and operating cost, it would have been much more difficult for nuclear energy to continue to be a significant part of the U.S. energy portfolio (ch. 1).[2]

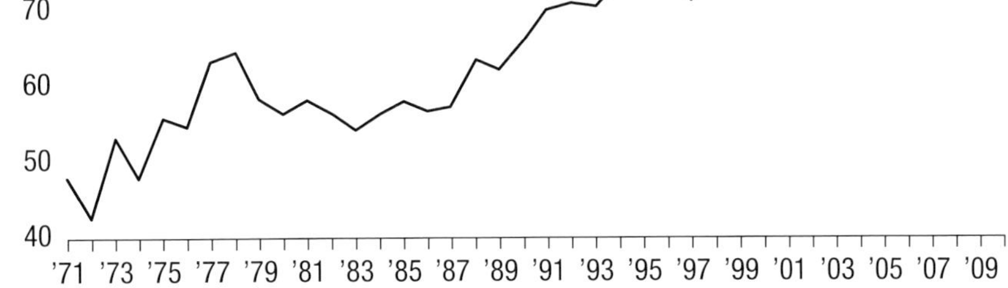

Fig. 8–1. U.S. nuclear industry capacity factors, 1971–2010, percent. (Source: Energy Information Administration, updated April 2011.)

In his influential guide to successful organizational change, John Kotter states that the driving force behind the process of transformational change is "leadership, leadership, and still more leadership."[3] Nuclear power's transformation in the electricity generation industry was no different: the driving force behind the industry-wide change was leadership. Today the industry's ability to sustain high levels of performance and continuously improve depends upon nuclear power leaders. The leaders that stepped up to transform the industry have played a crucial role in its survival. Many of these influential nuclear industry leaders have laid the groundwork for high standards and sustained improvements across the country. Other leaders served as role models for less-experienced nuclear professionals. Some nuclear leaders worked to develop strong leaders behind them and institutionalize the commitment to nuclear leadership development.

This chapter presents INPO's guidance on effective leader behaviors and goes on to examine what characteristics may be unique to nuclear power leaders. Initial research on the qualities of nuclear leaders is discussed, along with a sampling of industry leaders' viewpoints on nuclear power leadership. At the end of the chapter, nuclear leadership lessons that may be useful to other industries are presented.

INPO's Fundamental Attributes

A group of experienced industry leaders, together with organizational and leadership effectiveness subject matter experts, were asked by INPO to develop fundamental attributes for achieving and sustaining excellent nuclear station performance. The expert panel helped formulate the guidance document on the fundamental attributes and leader behaviors and actions that support strong station performance. The expert panel proposed six fundamental attributes of high-performing stations:

- **Core values.** Core values, including high standards of excellence and integrity, are clearly communicated, reflected in the behaviors of all station personnel, and modeled by leaders.
- **Vision and plan for excellence.** A compelling vision of excellence and a clear, integrated, and effectively communicated plan drive resource allocation, station priorities, and improvement.

- **Effective leadership team.** Leaders throughout the organization demonstrate and reinforce effective teamwork and collective ownership for results and the functioning of the organization. Collectively, leaders manage the human and technical aspects of the operation, focusing on continuous improvement, learning, and employee success.
- **Engaged employees.** Employees are valued, respected, and motivated and are committed to achieving excellence. The leadership team actively involves employees in problem solving and continuous improvement.
- **Healthy accountability.** Members of the workforce have the mind-set and capabilities to be accountable for demonstrating the core values, achieving results in support of the station vision and plan, and helping others do the same.
- **Effective processes and structures.** The organizational structure, administrative processes, staffing, and resource allocation support achievement of the mission, vision, and goals of the organization.[4]

To help illustrate how these fundamental attributes are demonstrated by leaders at the station, table 8–1 shows a selected subset of the leader behaviors and actions for each of the attributes.

The INPO guidance document is helpful in outlining what effective leaders do to achieve and sustain excellent station performance. It also includes a self-assessment for site leaders to identify gaps in behaviors and actions. The expert panel did not examine, however, the personality characteristics and competencies of effective nuclear leaders. Are there individual characteristics and competencies that make for a more effective nuclear power leader as well? Initial research on nuclear leaders examined the qualities or individual characteristics of effective nuclear leaders. Although the study primarily focused on nuclear supervisors, it offers potential insights into the characteristics defining nuclear leaders.

Table 8–1. INPO fundamental attributes of high-performing stations: Selected leader behaviors and actions

Fundamental Attributes	Leader Behaviors and Actions
Core Values	Regularly communicate defined organizational values and standards that reflect excellence. Visibly and passionately model the defined values and standards. Discuss management decisions in light of core values so that employees understand the "why" behind decisions.
Vision and Plan for Excellence	Establish a vision and define what excellence means in the areas of nuclear safety and plant reliability. Perform a rigorous assessment of current performance to ensure that the magnitude and depth of strengths and gaps to the vision and to excellence are understood. Maintain priority, alignment, and accountability for the execution of key improvement actions through formal and frequent reviews of progress and challenges.
Effective Leadership Team	Demonstrate shared ownership for improving both human and technical aspects of the whole organization, not just an area of specific responsibility. Facilitate collective reviews of operational events and conditions to stimulate teamwork. Collectively provide input for strategic planning and resource and staffing allocation, filling key positions, and prioritization of key activities.
Engaged Employees	Actively and visibly demonstrate respect, caring, and valuing of employees through acknowledgement, appreciation, and recognition. Communicate the bases for major decisions. When possible, provide a connection between the decision and the vision and values. Actively solicit and listen to employee input. Visibly follow up on identified concerns, and communicate actions taken to improve employee personal wellness, job performance, and relationships.
Healthy Accountability	Establish and consistently reinforce that coaching, feedback, and acting to correct performance shortfalls are expectations for workers. Openly acknowledge and be accountable for leadership shortfalls, and share the actions taken to correct mistakes. Provide ongoing feedback and coaching to reinforce adherence to defined standards and expectations.
Effective Processes and Structures	Establish and implement defined, structured planning and monitoring processes for core business functions and priority improvement initiatives. Establish clearly defined owners, roles, and responsibilities for the accomplishment of key station processes, including corrective action, work management, training, and self-assessment. Structure the organization to facilitate cross-functional communications, relationships, and responsibilities.

Initial Research on Nuclear Power Leaders

The research study set out to identify aspects of effective nuclear power first-line supervisors (ch. 10). Some of the findings will be discussed here because of their potential application to nuclear leaders in general. As part of the larger study, supervisors completed a leadership inventory that included a personality questionnaire. Supervisors were rated by their managers on a number of measures, including leadership competencies, and on their overall level of effectiveness as supervisors.

The results revealed that the first-line supervisors rated as more effective had the following personality characteristics: forthright and transparent, open to change and seeking ways to improve the status quo, group-oriented and affiliative, having low anxiety, and able to develop trusting relationships with others. Other leadership dimensions also found to be linked to effectiveness included having a sense of commitment to the organization and motivation to be a leader. Of the 19 competencies examined in the study, 8 were predictive of high performance in the supervisor (table 8–2).[5]

Table 8–2. Competencies most predictive of effective first-line supervisors

Inspiration	Motivates and energizes others to contribute at a high level
Drive for Results	Shows bias towards action and follow through on commitments
Problem Solving and Judgment	Identifies relevant information; makes timely and accurate decisions
Innovation	Challenges the status quo; looking to drive continuous improvement
Workload Management	Able to multitask; manages time effectively
Expertise	Demonstrates detailed knowledge and skill; keeps abreast of industry performance
Cultivating Relationships	Builds and maintains effective relationships
Strategic Vision	Has a vision for excellence; sees the big picture

Although these eight competencies were more predictive of effectiveness overall, both supervisors and managers rated two other competencies as most important by far: safety and integrity. Because these competencies were not found to be predictive of effectiveness yet were rated as most important, managers and supervisors were asked follow-up questions to help clarify the apparent discrepancy. Further discussions revealed that safety and integrity were seen as essential baseline or "price of admission" competencies for being in a supervisor role. Supervisors currently in the position demonstrated these competencies to some degree whether or not they were viewed as highly effective overall. Other aspects of the study having to do with nuclear first-line supervisor effectiveness, the results, and the conclusions are presented in chapter 10.

Implications

Upon reviewing the qualities of effective nuclear supervisors, it appears that they may share a number of characteristics with effective supervisors in other industries. These shared qualities and competencies may have to do with being group oriented, affiliative, developing relationships, and having the ability to motivate others and drive for results, as well as having good problem-solving skills and sound judgment, for example. More study is needed to verify whether nuclear FLS share some of the same characteristics with effective nonnuclear supervisors.

Many of the leader behaviors and actions supporting the six fundamental attributes of excellent plant performance may also be common across industries. A number of the leader behaviors and actions to support excellent nuclear station performance make for sound advice for any leader. "Perform a rigorous assessment of current performance to ensure that the magnitude and depth of strengths and gaps to the vision and to excellence are understood," for example, is a good first step for a leader of any large operation.

It is possible that in addition to the behaviors, qualities, and competencies that effective nuclear leaders have in common with leaders in general industry, there are other characteristics that stand out as vital and unique for nuclear power leaders. Compared to leaders in other industries, it is proposed that effective nuclear leaders are unique in the degree and manner in which they visibly demonstrate safety, integrity, and technical capability:

- **Safety.** An effective nuclear leader has high standards for safe behavior that are consistently demonstrated, while being an advocate for nuclear safety. The best nuclear leaders make sure that decision making reflects safety first, that communications are clear on safety over production, and that they consistently observe and coach others on safety standards. Nuclear leaders have to ensure that they are continually fostering nuclear safety culture by being approachable and open to input.

- **Integrity.** An effective nuclear leader demonstrates honesty and ethical conduct, in addition to a deep sense of responsibility for maintaining the trust and confidence of the public, the regulator, employees, corporation, community, and shareholders, by being transparent and self-critical. Nuclear leaders embrace their role as guardians of plant safety and of the trust that the public places in them. They see themselves as stewards of the technology to ensure no harm is done in the making of nuclear-generated electricity.
- **Technical capabilities.** Technical capabilities are crucial to an effective nuclear power leader. Effective nuclear power leaders are involved daily in technical decisions that have immediate and long-term implications. Nuclear leaders must possess technical competence in the operation of a nuclear power plant and the business of electricity generation.

Leaders' Perspectives

As a starting point for investigating the question of whether effective nuclear power leaders have unique characteristics as compared to effective leaders in other industries, a number of current and retired nuclear power leaders were interviewed for their perspectives on the subject. There are varied opinions among nuclear industry leaders themselves as to the ways in which nuclear power leaders mirror or differ from capable leaders in other industries. There is more consistency, however, regarding what nuclear leaders and others do to be successful.

Before Jack Davis became the CNO at DTE Energy, he was site VP at the North Anna nuclear station owned by Dominion and held several leadership positions in the U.S. nuclear navy. Davis emphasizes how nuclear leaders must maintain the trust and confidence of the public as well as their peers, given the potential impact of a disastrous event at any plant. "The plant is always under a microscope. And the leader needs to understand how the organization is viewed by all of its stakeholders in multiple venues. You must communicate with the public, the NRC, INPO, and people working at the plant, which is a level of regulatory oversight and public scrutiny not seen in other industries." Davis also states that it is important for nuclear leaders to be highly technically competent, which differs from other industries in which leaders may not need to have such deep knowledge of their technology, but they have solid management and leadership skills.[6]

Donna Jacobs is the VP of operations at Waterford 3 station, owned by Entergy. Prior to her current role, Jacobs was Entergy Nuclear's senior VP for planning, development, and oversight. She was previously at the Wolf Creek Nuclear Operating Company for much of her career, where she obtained her senior reactor operator certification. Jacobs's view regarding the similarities and differences among leaders is that "it comes down to risk. For those industries with a potentially hazardous impact on public safety, leaders need a high level of risk aversion and contingency planning. In nuclear power, it is ingrained in us from the beginning of our training to always ask, 'what's the worst thing that can happen?' For leaders in nuclear power and other industries, such as NASA, aviation, or the medical industry, public safety has to be an internal value." Jacobs also believes that nuclear leaders need to have a strong technical background to lead on a day-to-day basis and to make good decisions regarding long-term asset management.[7]

Prior to joining Arizona Public Service as the senior vice president of fossil generation, Mark Schiavoni was in a number of roles at Exelon, including president of Exelon Power and nuclear plant manager at the LaSalle County station. Schiavoni also served 22 years in the U.S. nuclear navy. Schiavoni suggests that for the most part, nuclear leaders do not differ from those in other industries. "The same traits are critical on both sides. The same desire to win, laser focus on what is important, to drive and not step back, to direct, and to be unafraid of difficult decisions. Excellent leaders also look outward to find out, 'How do I get better than that?' It's a 24/7 relentless pursuit." Schiavoni also talks about how excellent leaders are able to identify people who believe in what they are doing, those who can help lead the organization forward. Schiavoni adds, "With commercial nuclear power, there is nothing more important than protecting the public—so that comes with high levels of rigor around safety. Historically, however, in the fossil world, the feeling was, 'It's a tough job, there will be bumps and bruises, be a man about it.' So that is what we have to counteract as we improve safety performance in fossil generation."[8]

Joe Pollock is currently the executive director of nuclear operations at the Nuclear Energy Institute. Previously he was the site VP at the Indian Point Energy Center in Westchester County, New York, for four years, and has had a number of other leadership roles across the nuclear industry. According to Pollock, nuclear safety culture heavily impacts one's leadership style in nuclear power. "Our culture is to listen to workers, value their input and their questions, and you have to be very self-critical and transparent." Regarding the technical capabilities of nuclear leaders, Pollock points out that "the industry learned from the experience of the TMI accident that we need to have c-level leaders who have an understanding of commercial nuclear technology, which resulted in the emergence of the Chief Nuclear Officer role to oversee operations and advise the board."[9]

Oliver Kingsley served as the president of the World Association of Nuclear Operators (WANO) after retiring from Exelon as COO and president of Exelon Generation. Kingsley was the highly influential CNO of the ComEd, then Exelon, turnaround and the dramatic rise in performance across the industry. He has served as a director on boards in a variety industries. In Kingsley's view, successful nuclear leaders exemplify the same characteristics as successful leaders elsewhere. According to Kingsley, "Excellent leaders have core principles that they embody and exude, a very well-defined vision, and they are intolerant of things that do not work." Kingsley adds, "Leaders also need to put in place programs and processes, make sure workers understand their responsibilities and requirements, and have a way to assess how effectively programs and processes are being implemented. You need to have passion and a drive for excellence. Mostly, it takes guts."[10]

INPO's vice president of assistance, James Lynch, has served 30 years in various leadership roles at INPO and in the industry and has a master's degree in engineering from MIT. Lynch supports the hypothesis that nuclear leaders have unique characteristics, particularly regarding integrity. According to Lynch, "Integrity is remarkably high in the nuclear industry. So many nuclear executives started in the industry 30 years ago and know all about TMI and the founding of INPO. Nuclear leaders have a deep commitment to the nuclear industry. They are unfailingly honest, open, and self-revealing. The second distinguishing quality of nuclear leaders is the ability to drive for results, so they can produce electricity, with safety being the highest priority. It's a delicate balance and the best recognize how safety contributes to reliability." Being unfailingly honest, Lynch adds, "Bear in mind, however, that I have limited exposure to leaders in other industries."[11]

Eric Olson is the site VP at Entergy's River Bend Station in St. Francisville, Louisiana, where he previously served as the plant manager. Olson began his career in the nuclear navy on board a fast attack submarine. When he left the navy, he worked in multiple leadership roles in commercial nuclear power, many at Pilgrim nuclear station in Plymouth, Massachusetts. From the perspective of Olson, there are some important aspects of nuclear leaders in what they do to be effective. "Nuclear leaders need to understand the difference between high standards and actual standards as they get executed in the field. Then they have to put processes in place to continually evaluate that gap and fix it. Effective nuclear leaders also need to be skilled at developing the right leadership characteristics in others. Most importantly, they need to develop leadership in their team and develop the right behaviors in the organization."[12]

Tim Rausch, CNO at PPL's Susquehanna nuclear station in northeastern Pennsylvania, previously served as the site VP at Exelon's Oyster Creek station and fleet engineering director for Exelon, among other leadership roles in the industry. Rausch states, "Although there is some overlap with the characteristics of leaders in other industries, nuclear safety culture and safety conscious work environment have a unique impact on nuclear power leadership. You need to have a true appreciation for the people and the role they play in safe, reliable nuclear power generation—you cannot just trade them out quickly or efficiently." Rausch adds, "Trust and respect are also more critical as a nuclear leader. You need to have a relentless pursuit of continuous improvement such that excellence is barely acceptable. You have to check and adjust yourself as well. You have to lead consistently so your team can follow. If you're not consistent, and something goes wrong, it's harder to figure out how you got there."[13]

Transferring Nuclear Power Leadership Lessons to Other Industries

Although it may be that nuclear power leadership has some characteristics that are unique in how they are demonstrated (i.e., safety, integrity, and technical capability), there are nuclear industry leadership best practices that can be used by any industry where safety is a major concern while at the same time production demands are high. INPO's leader behaviors and actions supporting plant performance may be readily applied.

Another source of leadership best practices comes from the manager of organizational effectiveness at INPO. Bonnie Stahl's background is in human resources at Exelon and in leadership and organizational development at nonnuclear companies. She makes a number of observations of nuclear leaders that illustrate what they do at a behavioral level to build leadership alignment as well as operational focus:

- Communicate and reinforce goals, expectations, and values repeatedly with all employees, at all levels, and across multiple venues.
- Teach everyone how to be a leader, and expect everyone to demonstrate leadership. Teaching employees how to be a leader includes using events as opportunities and using multiple approaches to teach them specifically what leadership behaviors look like.
- Teach the maintenance shops how to allow for effective communications, and how to communicate about goals, expectations, and values on a regular basis.
- Use a common language across the site for providing feedback and to reinforce behaviors and expectations.
- When bringing on new employees, teach them about the expected behaviors and core values.
- Use standard meetings for practicing desired group and individual behaviors by deciding the purpose of the meeting and using an agenda and rigorous meeting processes.[14]

Although these actions are readily applied to general industry, Stahl emphasizes the importance of an effective human resources (HR) organization, to support leadership behavior and alignment. "HR infrastructure needs to support the long-term effectiveness of the operation. Those companies that do well have a strong HR function, where HR is a business partner with leadership."[15]

References

1 Walker, J. Samuel, and Thomas R. Wellock. 2010. *A Short History of Nuclear Regulation, 1946–2009*. Washington, DC: U.S. Nuclear Regulatory Commission. October.

2 *Nuclear News*. 2011. "U.S. Power Reactors to Be Examined for Vulnerabilities. May, 22–28.

3 Kotter, John P. 1996. *Leading Change*. Boston, MA: Harvard Business School Press.

4 Institute of Nuclear Power Operations. 2007. *Leadership Fundamentals to Achieve and Sustain Excellent Station Performance*. Atlanta, GA: INPO. September.

5 Rogers, Mary J., Ben K. Fearing, and Brian D. Juncker. 2011. "The Comprehensive Study of Nuclear First-Line Supervisor Effectiveness." *Strategic Talent Solutions*, June.

6 Davis, Jack. 2010. Interview with author. April 29.

7 Jacobs, Donna. 2012. Interview with author. March 26.

8 Schiavoni, Mark. 2010. Interview with author. August 19.

9 Pollock, Joseph. 2012. Interview with author. March 13.

10 Kingsley, Oliver. 2011. Interview with author. July 21.

11 Lynch, James. 2011. Interview with author. May 25.

12 Olson, Eric. 2011. Interview with author. October 21.

13 Rausch, Tim. 2012. Interview with author. April 26.

14 Stahl, Bonnie. 2011. Interview with author. October 13.

15 Ibid.

CHAPTER 9 The Turnaround Experts

Fifteen years ago when utility CEOs were looking to sell or close down their poor-performing nuclear power plants, people like Oliver Kingsley, now retired COO of Exelon, emerged to lead considerable turnarounds in plant performance. Kingsley and others transformed the troubled plants of Commonwealth Edison (ComEd/Unicom), merged them with the Philadelphia Electric (PECO) and AmerGen nuclear power plants, and created what is currently the largest and best-performing nuclear fleet in the industry, Exelon Nuclear.

Because Kingsley and other turnaround leaders tended to follow a pattern that was, and is, repeatable, the industry transformation built momentum. *Turnaround* refers to significant improvements made in the operation of a facility resulting in increased performance in safety, reliability, and production. As the turnaround experts moved on to other nuclear companies and applied their methods and skills to other stations in dire straits, they were explicit about their experiences and techniques. In line with the nuclear industry's culture of cooperation, these experts openly discussed and shared their lessons learned in public forums to help others benefit from their experiences.

The turnaround experts have made adjustments to their techniques over the years as they learned from the challenges of changing operations performance across multiple organizations. Not all nuclear leaders have promoted the same approach, and they personally demonstrate different styles. Nonetheless, there are commonalities and lessons learned that can be applied to other operations in need of a turnaround in safe, reliable production.

This chapter discusses the characteristics of the nuclear power turnaround leaders. Subsequently, it provides an overview of the tools used by the nuclear turnaround experts and how these tools can be applied to other organizations, nuclear and nonnuclear. This chapter starts with an illustration of the impact of one of the most influential nuclear power turnaround leaders, Oliver Kingsley.

Nuclear Turnaround Story

Shortly after Oliver Kingsley arrived at ComEd in 1997 as the chief nuclear officer, the new CEO John Rowe was concerned about the poor-performing nuclear power program and considered selling off nuclear assets.[1] ComEd was the Chicago-based owner of 12 reactors (at six sites) in the state of Illinois, which had spent billions of dollars constructing the units in the seventies and eighties over a long stretch, only to watch their performance deteriorate and their electricity production wither.

Multiple units within the ComEd Nuclear Generation Group (NGG) were under heavy scrutiny by the Nuclear Regulatory Commission (NRC). The LaSalle County, Dresden (in Morris, Illinois), and Zion stations were on the NRC's watch list, meaning they required additional oversight and had the potential of being shuttered by the NRC due to repeated regulatory violations. The units in the Quad Cities area appeared headed for the watch list as well. The NGG's capacity factor overall was dismal, safety performance was poor, and yet the company was spending millions every year to keep the stations operating. In addition, employee morale was low and labor-management relations were worse. In January 1997, the NRC issued a letter to ComEd expressing a lack of confidence in the company's ability to turn around the units. INPO was also highly critical of the company's culture and lack of improvement.[2]

When Mr. Kingsley said that the Nuclear Generation Group would be the best nuclear operator in the country, people did not believe him. Kingsley's vision seemed impossible. But the ComEd nuclear plants did get off the watch list (the board decided to shut down Zion). Units that were down were restarted, and within a few years were performing well, eventually breaking generation and safety records. Exelon Nuclear has been recognized as the best fleet of nuclear power plants in the United States and around the world.[3]

How did the ComEd/Exelon leadership team execute such a major turnaround? A review of the methods used by the industry's most effective turnaround leaders finds that nuclear turnaround experts principally relied on five tools to deliver sustainable turnarounds in overall operational and safety performance. Each of these tools is discussed below:

- Relentless and rigorous leadership at the top
- Focus on operations

- Align and engage the first-line supervisors, workforce
- Use processes to institutionalize improvements
- Check and adjust—people and the organization

Relentless and Rigorous Leadership at the Top

One of the first things Kingsley did at ComEd, then Exelon, was to evaluate the leadership talent throughout the organization to determine whether he had people with the right leadership behavior and skills. Kingsley exemplified the resolute leader at the top that became known for the nuclear turnaround. He was likely influenced, as the nuclear navy and nuclear industry have been, by the extraordinary Admiral Hyman Rickover, who was known for being as determined as he was demanding.[4]

When Kingsley started the turnaround at ComEd Nuclear, then Exelon, he evaluated and positioned leaders who had the drive, passion, and relentless pursuit of excellence necessary to lead the turnaround at the sites and in senior corporate roles. Kingsley had in his mind what it took to be an effective turnaround leader—how to lead the site and run the business. But like Rickover had done when bringing nuclear technology to the navy submarine, Kingsley depended on independent evaluations by management psychologists who knew the psychology of nuclear leadership.[5]

What they were looking for were individuals who were analytically sharp, clear-minded about the objectives, committed to the nuclear organization, and willing to do the hard work of the turnaround. Kingsley and his team put enough strong leaders with high standards in place to help them change the organization and fundamentally improve performance.

For another illustration of what the top nuclear turnaround leader does, we turn to the nuclear navy (retired) leader who led the notable turnaround at Fermi Unit 2, which is owned by DTE Energy in Detroit. After Jack Davis retired from the nuclear navy, he joined Dominion, where he earned his senior reactor operator's license and eventually became site VP at North Anna station in Louisa, Virginia. When Davis was hired as DTE's CNO in 2007, Fermi 2 had been on a steep decline despite a few years of good performance earlier in its lifetime. The site's feedback from their last

WANO assessment included numerous areas for improvement in operations, maintenance, equipment reliability, and supervisor effectiveness. In addition, the site was facing operational challenges daily.

While Davis assembled and aligned his leadership team, he told his organization, multiple times a day, through formal and informal meetings, that substandard performance was not acceptable and that they were capable of high levels of performance on par with the industry's best. With passion and emphasis, he described high standards and behaviors every day with each new situation. He told them what "great" looks like and what they would need to do to achieve it. He did not give up on making Fermi 2 one of the best sites in the nation, and he continually described excellence and insisted on it. Davis also obtained help in evaluating the individuals on the leadership team to help convert or recruit people who would exemplify the right leadership behavior. Gradually he was able to work through his managers on the senior leadership team to obtain higher standards and performance from the organization. According to Davis, "During a turnaround there is an inflection point where you become more facilitative and start to foster more debate and discussion. If you don't eventually allow the organization to run itself, it will become ineffective and people will leave." In Fermi's 2009 evaluation by INPO, they received acknowledgement of their operational improvements and a strength rating in organizational effectiveness.[6]

Diversity in the turnaround leader

Not everyone on the senior team needs to mirror the personal style of a Jack Davis or Oliver Kingsley. Not every successful turnaround leader does. Davis and Kingsley epitomize Rickover's nuclear navy–trained leader. What you do need are leaders with high standards, people who have a passion to make the plant better, who relentlessly challenge the organization to improve, and who make good use of the tools of organizational change at their disposal.

Tim Rausch is currently the CNO at PPL Susquehanna in Salem Township, Pennsylvania. Rausch was instrumental in the turnaround in performance and organizational health as the site VP at the Oyster Creek Generating Station. Rausch is an intelligent leader with a strong engineering background who took over at Oyster Creek on the heels of a rare nuclear plant 71-day union strike. With a composed demeanor, consistent approach, and high standards, he helped shape the Oyster Creek organization into a solid performer with a relatively supportive workforce that was open to the turnaround. He made sure he had an energetic plant manager and team of leaders that would raise standards and help him engage the organization.

Focus on Operations

Operational focus has such a crucial impact on the performance of a nuclear site that INPO has worked to develop a shared understanding of what the phrase means in the industry. *Operational focus* refers to the "behaviors and practices of an organization that are necessary to achieve high levels of operational safety and reliability."[7] The importance of operational focus at a nuclear plant seems obvious, and most plants maintain it to some degree, or they would not run well at all. But for a turnaround, operational focus is crucial. In a production facility with important safety concerns, such as the highly regulated environment of a nuclear power plant, other things can take center stage, including engineering challenges and debates, daily impacts on licensing requirements, maintenance schedules and problems, or even labor relations. For the station in the midst of a turnaround, the potential distractions are multiplied, in part from the assistance, concerns, and requests for information coming from the corporate organization, INPO, and the NRC. The turnaround experts know the significance of operational focus for getting the plant back on track and getting some run time.

Christopher Crane is an example of a demanding leader with high standards who continually shapes the organization to daily operational focus. Crane, who is now the president and CEO of Exelon Corporation, worked in his early years at Palo Verde, the large three-unit station located 45 miles west of Phoenix, Arizona. Oliver Kingsley recruited Crane to help lead the turnaround of the ComEd fleet. Kingsley knew Crane from their work together at TVA, where they had led site performance improvements starting in the early 1990s.

At ComEd and Exelon, Crane instituted standards, processes, and a leadership philosophy that helped drive each site to greater operational focus. According to Crane, "We needed to be more rigorous and so we put into place more management controls—policies, programs, and processes. Then we drove management controls while we created and nurtured the behavioral dimensions in order to sustain our focus and rigor."[8]

Chris Crane instituted standard operations-led mandatory meetings that focused the organization on the daily challenges to operations and helped others see that they needed to treat operations as the customer—that is, everyone else exists to support the safe, reliable operation of the nuclear unit. During these meetings, such as the daily morning call, Crane and the other senior leaders would question the site leadership in a manner that reinforced the critical importance of supporting safe and reliable operations.

Operational focus is particularly important in a turnaround as it aligns the organization around a common, cross-functional purpose, while simultaneously fostering safety and production (ch. 4). Crane and other turnaround leaders recognized the vital role operational focus plays in maintaining performance improvements.

Align and Engage the First-Line Supervisors, Workforce

Turnaround experts at first blush may appear to shortchange supervisor and workforce engagement. They do not overlook it. Oliver Kingsley individually interviewed 1,800 people during his first six weeks at ComEd and saw supervisor and workforce engagement as highly important. According to Kingsley, "Accept all types of feedback. What the worker thinks is a problem, take it seriously."[9] Tim Rausch believes, "You have to engage the employees in the turnaround and communicate each of the steps and why each is important."[10]

While they are reaching out to the workforce, turnaround experts also strategically approach employee engagement through alignment with the first-line supervisor (FLS), the fulcrum of influence at a nuclear generating station. A recent article in the *Harvard Business Review* illustrates how first-level managers can be crucial to a company's success.[11] Employee engagement survey results repeatedly show that workers have the most confidence in and alignment with their direct supervisors. The turnaround experts see the FLS as their most powerful ally, or potential enemy, in the battle to change how the organization performs. The FLS are also seen as critical members of the leadership team. Turnaround experts will often use a parallel process where communications are targeted at the workforce at the same time as they are engaging first-line supervisors. Turnaround leaders see that they must reset the standards, then enroll the first-line supervisors in improving the station and motivate them to engage the worker and change worker behaviors.

How to enroll the first-line supervisors in a turnaround

Turnaround leaders apply multifaceted interventions for enlisting the FLS in the turnaround. They pull the FLS into the leadership team by increasing communications and frequently hold regular group sessions with supervisors, managers, and directors together, referring to the meetings as *leadership alignment meetings*. To be most effective, these group sessions need to be held consistently, attendance should be mandatory, and the top leaders need to have heartfelt, honest discussions about the performance deficiencies and how the site is going to close the gaps. Leadership must avoid becoming "talking heads" who talk at the first-line supervisors in a one-way conversation.

The second part of enlisting the supervisor involves changing individual behavior and helps achieve the turnaround one person at a time. Some of the most successful turnarounds have used a leadership assessment and feedback process that gives leaders face-to-face conversations with each and every first-line supervisor at the station, plus every manager and director in between. The assessment and feedback process involves a cascading series of behavioral interviews that starts at the top of the organization and works down to the FLS. The interview is comprised of questions that directly tie to running the plant but with a focus on raising standards and increasing accountability.

Through the use of this intervention, the leaders know who is aligned with the standards and is eager to improve the station, who is on the fence and needs more coaching, and who has not really bought in or may not have the supervisory capabilities needed in the current environment. The facilitated feedback sessions provide straightforward coaching on the supervisor's leadership behavior, attitude, and alignment, including specific areas where they can improve.

The transformed and aligned supervisor is the voice on the ground for the leadership team and is expected to be the primary source for engaging the workforce in the turnaround. While top leadership sends out communications on the turnaround and holds all-employee meetings, the supervisor is critical to changing the behavior of workers and getting their input on problems and improvements on a daily basis.

Use Processes to Institutionalize Improvements

Chris Crane added another major component to the successful turnaround machine—put the best methodologies into a process or procedure in order to institutionalize the right approaches. At the height of the turnaround wave, most nuclear plants had not yet assumed the highly rule-based, procedure-driven approach that Exelon has adopted, which is now more widespread in commercial nuclear power production. Before the shift to process and procedure, plant activities relied heavily on the knowledge and coordinated relationships of the highly tenured employees at each station. Crane made "institutionalize" common verbiage in the nuclear power industry and developed the most highly process-driven, and successful, nuclear power company. This put Exelon ahead of the game when attrition due to retirements started worrying the industry and some nuclear operations were still heavily dependent on personal, historic knowledge.

The reason why process improvement and proceduralizing are important in a turnaround is that they make expectations clear and concrete and help formalize a course for improvement efforts. Find out who does something the best, figure out how to do it at your plant, and then have everyone do it that way every time. Benchmarking and formalizing processes sound simple, but even nuclear organizations differ in their tolerance for proceduralizing work. INPO has acknowledged that effective organizations vary in how heavily they are rule-based. However, sufficient and effective process and structure are fundamental for achieving and sustaining excellent station performance.[12]

For turnarounds, becoming more formal, consistent, and process-driven is crucial. Most operations that are in need of a turnaround conduct their business too casually, without rigor, and lacking adherence to high standards, or even to their own procedures. Finding the best practices and putting them into formal processes or procedures provide a success path for the site.

One of the best practices is to systematically evaluate your processes and regularly upgrade them. Rather than become wedded to what and who worked in the past, or becoming victim to the "flavor of the month," turnaround experts continually monitor how things are going and make adjustments continually. This practice is essential for sustainability. After turnarounds gain some traction, performance often drops off again, and plants become cyclic performers after the people breathe a sigh of relief, as if to say, "Thank goodness that's over." Continuous improvement saves the turnaround from complacency, prevents the organization from suffering the leader who falls in love with his own ideas, and keeps the organization open to new ways and innovations, all of which are staples of the healthy nuclear operational learning culture. The turnaround experts also put into place "people processes" that continually evaluate and upgrade leadership capability as well.

Check and Adjust—People and the Organization

The fifth tool in the successful nuclear turnaround is continually evaluating and improving your current and future leadership and the organization that supports them. On face value, talent assessment and organizational development do not sound like turnaround methods, but successful turnarounds that sustain improved performance include a way to do a check and adjust on their people. By evaluating the current and potential leadership talent, top leadership can calibrate attitude and alignment as well as determine whether they have enough people with the right leadership behavior to meet current and future goals. By "dip-sticking" the organization with targeted tools, it is possible to monitor and track trends in morale, alignment, and any issues particular to the turnaround.

The best nuclear operators use talent reviews as a turnaround device to build alignment regarding the standards and expectations for leadership behavior in the organization. In order to check and adjust people in ways that drive improvements and sustain the turnaround, the tools need to be designed to work in an environment of high demands for safety and reliable production.

By defining behavior, gathering data, and discussing people together, the leadership team builds a shared understanding of what it is going to take to turn around the organization and continue improving. By regularly providing individuals assessment-based feedback on behavior, more people are pulled in the right direction. The individual conversations also allow for feedback up the chain, as well as provide key information for succession planning.

According to Chris Crane, "You need talent management tools to identify individual behavioral dimensions and proficiency. But you need to effectively execute your talent management programs and processes as you would your equipment reliability programs and processes."[13]

For the organization-level analyses, tools designed for the safe-production, high-intensity environment provide the most valid and useful information. In turnarounds, some level of turmoil and consternation are to be expected and are seen as initial signs of organizational change. Top leadership can minimize the impact of potentially distracting and negative trends in morale by tracking organizational issues and developing responses, adjustments, and communications as needed.

Organizational checks should be both formal and informal. Formal methods, such as employee surveys, need to be tailored and not just administered on autopilot, or they tend to wear people out, and the information becomes less useful. Informal tools can take many forms. One senior vice president of operations at one point directed all his site vice presidents to develop a variety of methods to track people issues and reactions and communicate about potential problems with organizational change efforts. The various methods included "fireside chats," small employee group advisory councils, breakfast with the site VP, and skip-level meetings, as well as all-employee and first-line supervisor meetings. These regular check-ins can be used to complement employee survey data and organizational and senior team analyses as they also provide a venue for discussion. To be effective, issues that are raised must receive honest responses from leadership and follow-up actions must be tracked and communicated.

Transferring Nuclear Turnaround Tools to Other Industries

While it is true that the nuclear power industry is unique, nuclear power plants are essentially production facilities that must function with exceptional safety and high reliability. They are not unlike many other generation or production facilities, especially those with high safety and performance demands. The new general manager brought in by senior management to get things turned around at a struggling facility can readily apply the lessons learned from the nuclear turnaround experts starting with the five turnaround tools. Table 9–1 outlines how the five essential nuclear turnaround tools can be leveraged in a nonnuclear environment.

Table 9–1. Turnaround tools and interventions

Turnaround Tools	Interventions
Relentless, Rigorous Top Leadership	Put in place a leader with high expectations for performance and quality, someone who is inspiring and demanding; evaluate and recruit a leadership team comprised of people who have high standards and are able to engage the organization in change.
Operational Focus	Drive the organization to focus on critical operations with regular status meetings and daily discussions of challenges to operations; use operational focus as a unifying force to build cross-functional teamwork that resolves long-standing problems.
Align and Engage Supervisors, Workforce	Align and engage the FLS through targeted interviews and feedback; use group and individual interventions to reinforce expectations and enroll FLS in improvements; enlist the FLS to help engage and align the workforce; communicate regularly with the workforce on changes, gather input, and respond.
Use Processes to Institutionalize Improvements	Develop written directions based on best practices; define and record processes/deposit tribal knowledge into the process for how work gets done; make regular self-assessments of the processes and best practices, and make improvements.
Check and Adjust—People and the Organization	Set up regular evaluations and team reviews of key management positions to determine fit with current conditions and anticipated needs; provide frequent feedback; use formal and informal ways to track organizational trends, concerns, and progress while communicating frequently on results.

Another resource for turning around performance in an organization is John Kotter's best-selling book, *Leading Change*.[14] Kotter thoroughly describes how to lead an organization through transformational change and the many pitfalls to avoid. In 2005, INPO created a guidance document on the eight steps for improving nuclear plant performance, which was based on Kotter's book and modified for nuclear power plants.[15] Figure 9–1 shows Kotter's eight-stage process in its original form, which is useful for any industry.

1 Establishing a Sense of Urgency

- Examining the market and competitive realities
- Identifying and discussing crises, potential crises, or major opportunities

↓

2 Creating the Guiding Coalition

- Putting together a group with enough power to lead the change
- Getting the group to work together like a team

↓

3 Developing a Vision and Strategy

- Creating a vision to help direct the change effort
- Developing strategies for achieving that vision

↓

4 Communicating the Change Vision

- Using every vehicle possible to constantly communicate the new vision and strategies
- Having the guiding coalition role model the behavior expected of employees

↓

5 Empowering Broad-Based Action

- Getting rid of obstacles
- Changing systems or structures that undermine the change vision
- Encouraging risk taking and nontraditional ideas, activities, and actions

↓

6 Generating Short-Term Wins

- Planning for visible improvements in performance, or "wins"
- Creating those wins
- Visibly recognizing and rewarding people who made the wins possible

↓

7 Consolidating Gains and Producing More Change

- Using increased credibility to change all systems, structures, and policies that don't fit together and don't fit the transformation vision
- Hiring, promoting, and developing people who can implement the change vision
- Reinvigorating the process with new projects, themes, and change agents

↓

8 Anchoring New Approaches in the Culture

- Creating better performance through customer- and productivity-oriented behavior, more and better leadership, and more effective management
- Articulating the connections between new behaviors and organizational success
- Developing means to ensure leadership development and succession

Fig. 9–1. Kotter's eight-stage process of creating major change. (Source: Adapted from Kotter, John P. 1995. "Why Transformation Efforts Fail." *Harvard Business Review*, March–April: 61. Reprinted with permission.)

References

1 *Bloomberg Businessweek*. 2010. "John Rowe's Bet on Nuclear Energy." September 23.

2 Washington International Energy Group. 1998. *Need for Natural Gas Increases with More Nuclear Plants Shut Down*. Washington, DC: INGAA Foundation, Inc. http://www.ingaa.org/File.aspx?id=540.

3 Kingsley, Oliver. 2011. Interview with author. July 21.

4 Rockwell, Theodore. 2002. *The Rickover Effect*. Lincoln, NE: iUniverse.

5 Mahaffey, James. 2009. *Atomic Awakening*. New York: Pegasus Books. 273.

6 Davis, Jack. 2010. Interview with author. April 29.

7 Institute of Nuclear Power Operations. 2010. *Principles for a Strong Plant Operational Focus (Preliminary)*. Atlanta, GA: INPO. July.

8 Crane, Christopher. 2010. Interview with author. August 6.

9 Kingsley, Oliver. 2011. Interview with author. July 21.

10 Rausch, Tim. 2012. Interview with author. April 26.

11 Hassan, Fred. 2011. "The Frontline Advantage." *Harvard Business Review*, May, 106–114.

12 Institute of Nuclear Power Operations. 2007. *Leadership Fundamentals to Achieve and Sustain Excellent Station Performance*. Atlanta, GA: INPO. September.

13 Crane, Christopher. 2010. Interview with author. August 6.

14 Kotter, John P. 1996. *Leading Change*. Boston, MA: Harvard Business School Press.

15 Institute of Nuclear Power Operations. 2005. *Template for Significantly Improving Nuclear Plant Performance*. Atlanta, GA: INPO. December.

CHAPTER 10 The Nuclear First-Line Supervisor

Industry leaders readily acknowledge that first-line supervisors have some of the toughest jobs and play a vital role in the success of the operation. A recent *Harvard Business Review* article illustrated how the first-line supervisors (FLS) can be crucial to a company's success, especially during a turnaround. In the case study described in the article, the company CEO singled out the FLS to help mobilize the broader organization.[1] Typically the FLS are caught in the middle, where they have demands from managers above to get things done right, as well as pressure from workers, who present concerns, questions, and potential barriers to getting work done that need to be addressed. First-line supervisors working in the context of a nuclear power plant have a unique overlay. Like other supervisors at production sites, nuclear supervisors have to get workers out in the field doing the right work—and doing the work right every day. Unlike many other industrial settings, at a nuclear site, safety concerns are sacrosanct. The nuclear first line has to support a safety conscious work environment (ch. 3) in which employees are encouraged to raise safety concerns to a responsive management.[2] While getting work accomplished, the nuclear supervisor is expected to keep safety—nuclear, radiological, and industrial—the primary goal at all times. If workers do raise any safety concerns or questions, the FLS is required to respond appropriately and address the issues promptly, including directing the worker to submit the issue to the corrective action program.

The nuclear power FLS also has a significant role and substantial responsibilities as part of the broader management and leadership team at the site. As an extension of management, nuclear supervisors support continuous improvement by identifying conditions adverse to quality and completing corrective actions. They recognize training needs in their organizations while making sure workers' qualifications are up-to-date. Nuclear FLS also review and make improvements to work packages in addition to driving the daily work schedule. In managing their people, supervisors ensure that workers comply with human performance error prevention expectations, along with safety standards, federal regulations, and company policies. They observe, coach, develop, and performance manage workers, represent management in meetings, and communicate management's messages to workers.

This chapter examines the demanding role of the nuclear first-line supervisor by presenting a two-part research study validating industry guidance and expanding the understanding of what makes supervisors effective. At the conclusion of this chapter, applications to other industries are presented.

Background

In the past decade, nuclear power industry leaders began to recognize the critical importance of first-line supervisors and their impact on how well the station was performing. Despite substantial gains in plant performance across most of the industry, some stations had continued to struggle with errors and equipment problems that led to substandard results. INPO's analyses of plant performance data found that poor supervisor performance was frequently the cause of, or a contributor to, plant events and problems. INPO subsequently formed a working group on the nuclear power plant FLS, which led to published guidelines for effective nuclear supervisor performance.[3]

The nuclear FLS working group was made up of people from INPO and the industry with experience in management and supervisory positions at nuclear power plants. The working group identified core attributes, management and leadership abilities, and roles and responsibilities that make for an effective FLS.

Presented below are the core attributes of the nuclear FLS that the working group identified, with selected examples of specific characteristics for each attribute:

- **Has uncompromising high standards and integrity.** Has high expectations for self and others; hold workers, peers, and those senior in the organization accountable.
- **Is willing to do the hard work of critical thinking.** Influences group discussions with sound thinking; recognizes issues early, thinks through and confronts implications.

- **Accepts responsibility.** Takes ownership of personal and subordinate mistakes and learns from them; aligns with and communicates management decisions with ownership.
- **Maintains an upbeat, positive outlook.** Displays a deep-seated passion for the business; displays eagerness to participate and learn; lifts the collective mood in group settings.
- **Displays a pattern of good judgment.** Gathers facts, listens intently, is open to hearing problems, and is consistent in responding to problems; advocates and ensures nuclear, radiological, and personnel safety; consults and collaborates with others.
- **Is willing to deal with difficult employee issues.** Has courage to professionally confront inappropriate employee behavior; identifies employee problems early and works to resolve them.[4]

The guidance document includes a thorough exposition on the attributes of nuclear supervisors and their roles and responsibilities. Also included are aspects of FLS alignment with management as well as suggestions for the selection and training of FLS. The only missing element was basic research to validate the working group's findings regarding effective nuclear supervisors and the important role of alignment with senior management.

Nuclear First-Line Supervisor Research

Since the emergence of the supervisor guidelines, the industry has been more selective in putting people in the FLS position and more attentive to how well incumbents are aligned with leadership's expectations. Strategic Talent Solutions (STS), a management consulting partnership based in Chicago, has had the opportunity over the past few years to evaluate and interview hundreds of nuclear supervisors in the United States and England. STS has found that there are some individuals who are remarkably effective and embrace the weighty role of the nuclear first-line supervisor. Then there are others who struggle miserably and become frustrated and bitter in the role. There are also many nuclear supervisors who are earnest and capable and do a decent job but could have a larger impact on improving station performance.

Because supervisors can heavily influence the trajectory of a nuclear power station, STS set out to validate INPO's guidelines and determine whether there are any other factors that make for a highly effective nuclear supervisor. Some of the research questions included, "Are there unique characteristics or motivations that are predictive of the high-performing FLS? What do effective supervisors do that others fail to do? What can be done to help supervisors perform better?"

A two-part research study was conducted to answer these questions. Part one involved a job analysis to verify actual responsibilities and competencies, while the second part of the study examined the experience, training, alignment, motivations, and other characteristics of the high-performing nuclear FLS.[5] The results of the study and the answers to these questions have implications for selecting and developing FLS in any industrial production environment where safety is a significant concern.

Job analysis

In order to ascertain the characteristics and behaviors of the effective nuclear supervisor, the study first defined the actual responsibilities and tasks of the nuclear supervisor. A job analysis was conducted to determine major areas of responsibility, specific tasks, competencies, and perceptions regarding organizational alignment of the nuclear FLS.

The sample for the job analysis included 119 nuclear supervisors and 54 managers from three different nuclear stations (and two different companies) who completed a series of questions about the FLS position. Participants rated items about tasks and responsibilities in terms of priority, frequency, and importance.

Responsibilities and tasks. The study included an analysis of 57 specific tasks of the nuclear supervisor. These tasks were grouped into seven major categories. As shown in figure 10–1, supervisors spend the majority of their time scheduling and assigning work (22%) and supervising work (24%). When combined with scheduling and assigning *emergent* work and coaching their direct reports, respectively, these two broad areas capture the vast majority of their time: scheduling and assigning work (39%) and supervision and coaching of direct reports (36%).

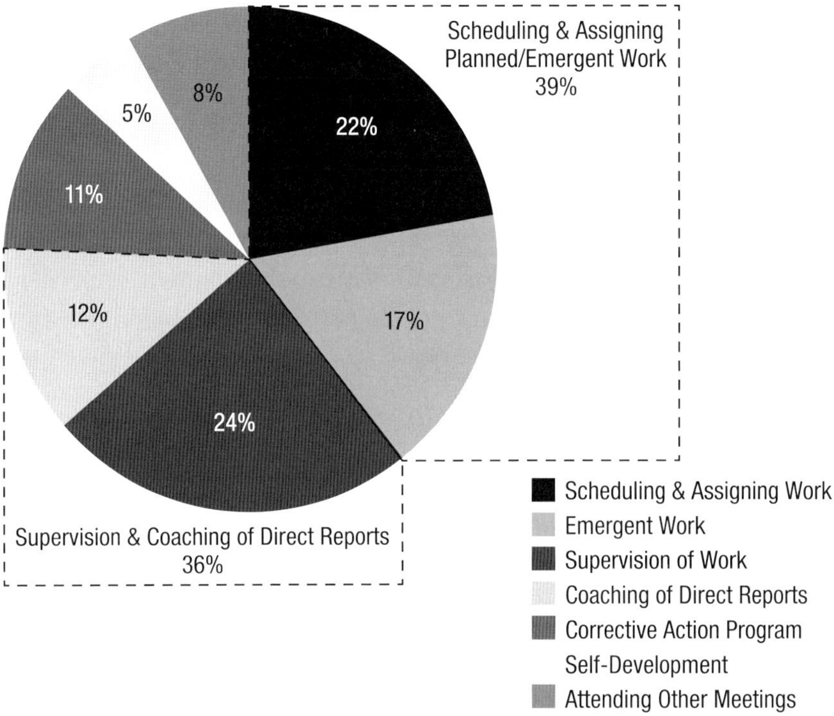

Fig. 10–1. Percentage of time FLS spent on major responsibilities

Ratings of the priority of responsibilities were consistent with how supervisors report spending the majority of their time. Moreover, supervisors and managers showed agreement when they were asked to rank the priority of FLS's areas of responsibility (from highest to lowest priority):

- Supervision of work done by direct reports
- Providing feedback, correcting behaviors, and coaching
- Scheduling and assigning work
- Managing emergent work and issues
- Responsibilities related to the corrective action program
- Self-development and continuing education
- Attending meetings unrelated to the areas listed

Task frequency and importance. Task frequency and importance ratings were consistent with the major areas of responsibility results. Those tasks with the highest frequency and importance ratings had to do with overseeing direct reports in multiple facets and assigning and scheduling work. The top four tasks rated the most important pertained to identifying and correcting worker safety violations and addressing worker substandard performance.

Administrative burden. It was noted that "administrative burden" is a common complaint of supervisors and was mentioned frequently in interviews. Follow-up first-line supervisor observations were conducted to clarify what kinds of tasks comprise administrative burden according to supervisors. The observations revealed that the majority of administrative tasks that the FLS were referring to included actions having to do with fixing work packages, getting on the computer to manage work packages, and actions on the computer related to the corrective action program. Hence, computer work was categorized according to the area it supported. Meetings were similarly categorized and included an "other" meeting category.

Supervisor competencies. The INPO guidelines on effective supervisor attributes (presented previously) and behaviors were parsed into distinct concepts, which formed a total of 19 unique supervisor competencies. Participants in the job analysis rated the importance of the 19 competencies for executing the role of FLS effectively.

Results showed that the attributes and behaviors cited as important for supervisor effectiveness by INPO were also identified by respondents in the study as important competencies for effectiveness in the role of the supervisor. Safety and integrity had the highest ratings. Safety was rated as critically important, the highest category, by 80% of respondents. Other competencies rated highly important included working through problems and making good decisions, setting high standards, and holding themselves and others accountable.

Organizational alignment and support. The study also examined aspects of alignment and perceived management support of the FLS. Supervisors and managers were asked for their level of agreement with 21 statements, such as "Supervisors are treated as core members of the management team." Notably, most (80%) reported that supervisors are both responsible for aligning the organization around standards and for providing solutions and corrective actions to improve the plant. The majority (75%) also reported that supervisors know what is expected from them on the job. Although the vast majority (90%) stated that supervisors are encouraged to provide on-the-spot performance feedback, only 25% stated that FLS are rewarded for coaching and mentoring others. A minority (36%) suggested that support for the transition to the management team was sufficient. Few (22%) thought FLS were effective at balancing administrative responsibilities with time in the field. It was striking how few FLS reported that the supervisor position is a desirable role (16%).

Manager and supervisor differences. Specific aspects of organizational alignment and management support were the only areas where differences were found between managers and supervisors on the role of the supervisor in the job analysis. Of the 21 items, 13 showed manager-supervisor differences of 5% or greater (fig. 10–2). For instance, more supervisors (76%) than managers (54%) indicated that FLS are rewarded for getting work done, and supervisors were less inclined than managers to say that FLS were treated as core members of the management team (61% vs. 71%).

Conclusions. The results of the job analysis clearly support the INPO guidelines in terms of the core attributes of the nuclear supervisor and primary responsibilities. No significant differences were found between supervisors' and managers' perceptions of the FLS's tasks (frequency and importance), time spent on major areas of responsibility, and importance ratings of supervisor competencies. They differed, however, on a number of organizational alignment items. Some of the differences pertained to perceptions of support or backing of the FLS and the degree to which FLS feel like they are treated as core members of the management team.[6]

An important question remains, however: How important are the competencies, time spent on responsibilities, and management support or alignment in predicting high levels of FLS effectiveness? Part two of the study set out to answer this question and obtain a deeper understanding of the successful nuclear FLS in order to identify ways to help supervisors be more effective.

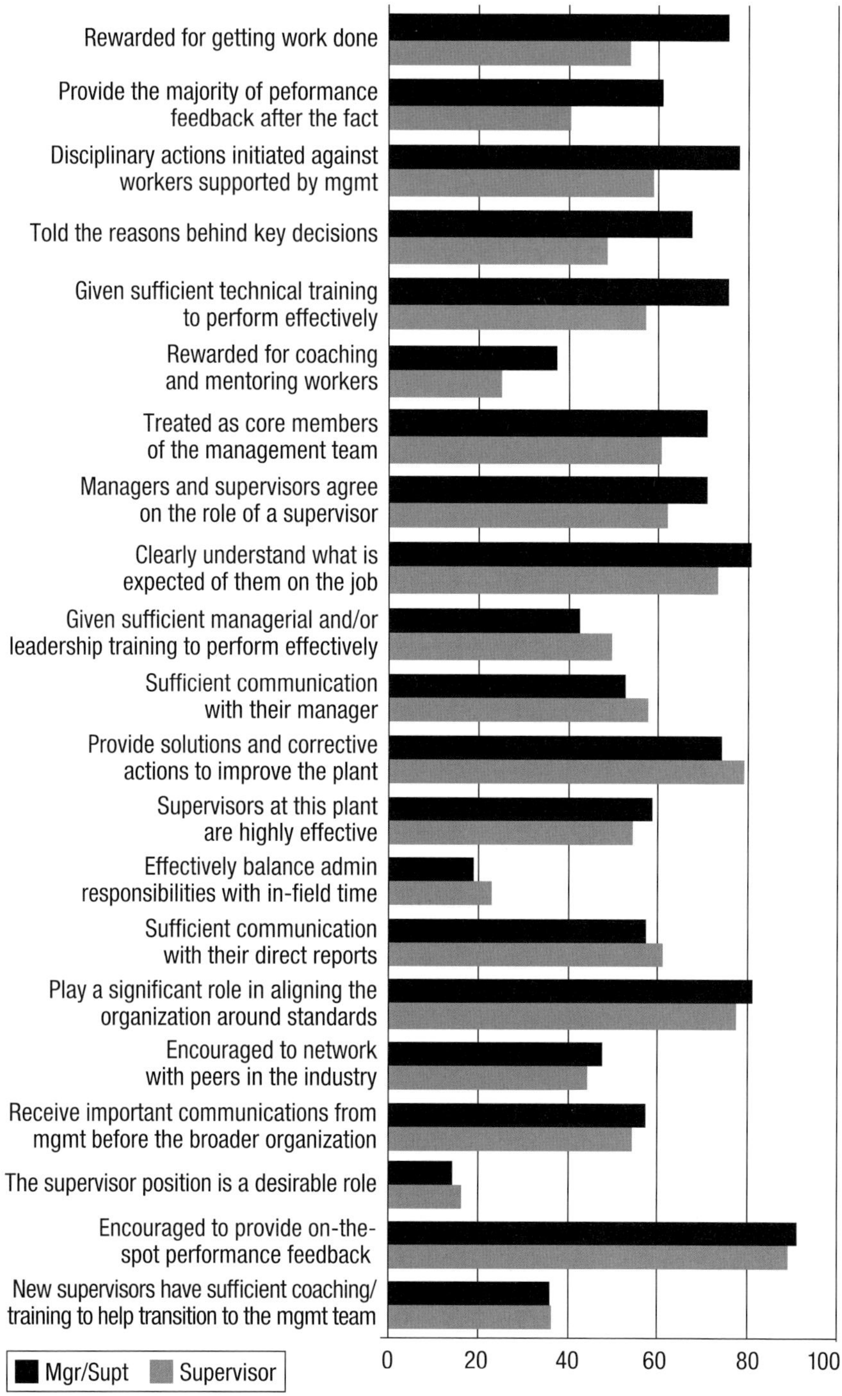

Fig. 10–2. Manager and supervisor alignment ratings

Predicting nuclear first-line supervisor effectiveness

A larger sample was used for part two of the study: 281 supervisors and their managers from four different nuclear sites (at three different companies) participated. Supervisors completed a series of online questionnaires about their experience, training, and time allocation, as well as questions to assess their motivations, confidence, job fit, role clarity, organizational commitment, and alignment. Managers provided ratings on competencies, overall effectiveness, and transition to management for the supervisors who participated.

The analyses examined whether manager ratings of supervisors were related to the following variables:

- Education and experience (including military)
- How supervisors spend their time at work
- Training (including INPO-delivered training)
- Organizational commitment and alignment
- Leadership dimensions (competencies, motivations, fit, confidence, role clarity)[7]

Supervisor education and experience. No significant relationships were found between education level and manager ratings of effectiveness. There were differences, however, relating to years of supervisory experience, military background, and outage experience. First-line supervisors with more supervisory experience were rated more effective. They also reported having a better understanding of the requirements of the role, having made greater progress transitioning to the management team, and having more confidence in their leadership capabilities.

Those with military experience were less likely to feel they were treated as core members of the management team, that they received communications in a timely manner, or were involved in management decisions. The number of refueling outages during which an individual had been a supervisor was correlated with high manager ratings of performance and progress in transitioning to the management team.

How the FLS balance their time. Supervisors who reported spending a greater percentage of their day providing oversight to their direct reports were rated by their managers as more effective overall and better transitioned to the management team. Supervisors who report spending more time managing emergent work and emergent issues were rated by their managers as being lower in overall effectiveness.

Supervisor training. Most supervisors reported having received supervisor or management skills training (88%). Moreover, those who had received INPO supervisory training (33%) tended to consider themselves, and were rated by their managers as, being further along in the transition to the management team. They were also more likely to report that they had received sufficient managerial and/or leadership training. Conversely, those supervisors who reported that inadequate training had made it more difficult to do their job were also rated as less effective and as having made less progress transitioning to the management team.

Organizational alignment. Supervisors who had made greater progress in transitioning to the management team were viewed as more effective overall by their managers. Organizations play a role in assisting supervisors in making this transition in various ways. Supervisors who report that they are treated as core members of the management team and that they have sufficient communication and personal contact time with their managers are significantly more likely to have made the transition to the management team, and are rated as more highly effective.

Leadership dimensions. The analyses of the leadership dimensions showed that manager ratings of effectiveness were higher for those supervisors who prefer to assume leadership positions, are confident in their ability to lead others, believe they are well-suited to the position, express a commitment to the organization, and clearly understand what is expected of them. Ratings of supervisors on specific competencies were also related to manager ratings of effectiveness. That is, supervisors seen as highly effective by their managers were also highly rated on motivating others, drive for results, problem solving and judgment, innovation, and personal workload management.

Summary and implications

The picture of nuclear supervisor effectiveness is significantly more detailed given the results of the two-part study. Moreover, there are actions management can take to improve FLS alignment and performance.

In summary, highly effective nuclear supervisors:

- See themselves as leaders and as well-suited for their position.
- Report that they are treated as core members of the management team. Their managers indicate they have made good progress in their transition to management.
- Have a clear understanding of expectations and are committed to the organization.
- Treat safety as the highest priority; demonstrate integrity.
- Are accountable, good problem-solvers, have a drive for results, motivate others, and build relationships.
- Have attended leadership training, including the INPO supervisory course.
- Spend more of their time providing oversight and coaching to direct reports relative to assigning work and pursuing emergent issues.[8]

The findings have implications for selecting well-suited individuals but also for on-boarding, training, and development opportunities. Management can have an impact on supervisors in the steps they take to get new supervisors on board, build alignment, and assist supervisors with the transition to management. They can improve the transition by treating supervisors like core members of the management team. That encompasses communicating proactively with first-line supervisors, involving them in decisions, sharing the reasons behind decisions, and by providing sufficient communication and personal contact time.

Translating Nuclear FLS Lessons to Other Industries

These supervisor results apply to industrial environments where both production and safety are necessary to high-performing operations. Table 10–1 summarizes the findings regarding effective FLS and the steps general managers (nuclear and nonnuclear) can take to leverage the capabilities and potential power of the FLS. The starting point is appreciating the extent to which FLS performance can impact plant performance, followed by proactively and regularly communicating with supervisors and building their alignment with the management team. Plant general managers should make a simple plan to leverage their FLS group to help improve station performance and get their team involved in the discussions.

Table 10–1. First-line supervisor effectiveness: Applications to general industry

Topic	Research Findings	Recommended Actions
Supervisor Selection	Specific leadership dimensions predict supervisor effectiveness: Confidence in leadership role and skills, commitment, integrity, problem solving and judgment, accountability, inspiration, and building relationships.	Use a validated selection process designed for supervisors in high safety/production environments and linked to behavioral dimensions predictive of performance in the role.
New Supervisors	Supervisors with high-quality training, who gain alignment with senior management, perform better.	Develop a supervisor on-boarding process that includes frequent communication and personal contact time with managers. Use training that is targeted to the role and builds alignment with management for greater impact.
Military Experience	Supervisors with military backgrounds are less likely to feel they are core members of the management team; alignment with management is predictive of better performance.	Ensure that military recruits have sufficient on-boarding experiences that build alignment. Use mentors who have successfully made the cultural transition to management.
Struggling Supervisor	Supervisors who spend more time on emergent issues than directing and coaching their people perform more poorly, as do those with less confidence and who have made a slower transition to the management team.	Evaluate struggling supervisors for fit with the position and to assess the behavioral gaps to better performance, including how they balance their time. Make a plan to close gaps with high involvement of the manager.
Low-Performing Supervisor Group	Higher performing FLS clearly understand expectations, see themselves as core members of the management team, are committed to the organization, accountable, good problem solvers, have a drive for results, motivate others, and build relationships.	Conduct a two-part intervention: (a) Increase alignment via proactive communications and engagement with the FLS group, increase contact time with managers to involve FLS in decisions, explain the reasons behind decisions, and clarify expectations. (b) Evaluate supervisors individually for fit with the position and gaps in leadership behavior. Formalize plans to close behavioral gaps; move those with poor fit to other positions.

References

1 Hassan, Fred. 2011. "The Frontline Advantage." *Harvard Business Review*, May, 106–114.

2 U.S. Nuclear Regulatory Commission. *NRC Regulatory Issue Summary 2004-xx Guidance for Establishing and Maintaining a Safety Conscious Work Environment*. Washington, DC: U.S. Nuclear Regulatory Commission. http://www.nrc.gov/about-nrc/regulatory/allegations/scwe-guide.html.

3 Institute of Nuclear Power Operations. 2007. *Guidelines for Effective Nuclear Supervisor Performance*. Atlanta, GA: INPO. November.

4 Ibid. 3.

5 Rogers, Mary J., Ben K. Fearing, and Brian D. Juncker. 2011. "The Comprehensive Study of Nuclear First-Line Supervisor Effectiveness." *Strategic Talent Solutions*, June.

6 Ibid.

7 Fearing, Ben K., Brian D. Juncker, and Mary J. Rogers. 2012. "Selecting and Developing Effective First-Line Supervisors in the Nuclear Power Industry." *Strategic Talent Solutions*, March.

8 Ibid.

CHAPTER 11

Site Senior Leadership Team

Commercial nuclear power site senior leadership teams face considerable challenges to keep their operating units running safely and reliably throughout the entire fuel cycle. Research and experiences working with nuclear power and other production teams show that the strength of the senior leadership team (SLT) is crucial to achieving and sustaining high performance. The effective SLT faces these challenges together in a way that makes for better decisions and better results. Moreover, there are relatively simple steps leaders can take to improve the functioning of the team and its ability to lead the station to perform.[1]

In many ways, the nuclear power plant site provides a good environment for studying team effectiveness. The nuclear SLT is a true "team" in that the members have complementary roles, are committed to a common purpose—plant performance—and are held accountable for shared results.[2] The nuclear leadership team typically includes the top 12 or more people at the station who are responsible for the daily operation of the plant but also have an impact on long-term asset management. The nuclear power plant SLT is made up of the site's top leaders, who typically are the site VP and plant manager, plus the heads of operations, engineering, maintenance, work control, radiation protection, chemistry, regulatory assurance/licensing, training, human resources, finance, and nuclear oversight or quality assurance.

Although SLT members have significant direct accountabilities for their individual departments, consistently high station performance is dependent upon the team working well together. The effective running of a nuclear site demands such teamwork in part because of the continual internal and external demands related to safety, regulation and licensing, production, and cost. The interdependence of the department functions to manage all these pressures on a daily, as well as strategic, basis makes effective teamwork essential to good plant performance.

This chapter discusses the characteristics of effective nuclear power SLTs and provides an illustration of the work of one team through an evaluation, intervention, and effectiveness review. Applications to senior leadership teams outside of nuclear power are provided at the conclusion of this chapter.

What Makes for Effective Teamwork?

INPO includes the effective leadership team as one of the fundamental attributes of high-performing nuclear stations.[3] "Effective teamwork" can sound a bit like motherhood and apple pie because most people feel that they know good teamwork when they see it. Leaders can recall times when they have been on effective and ineffective teams and will juggle people around or take people out on activities together to try to improve team functionality. What leaders often fail to recognize is that they may lack the ability to objectively and accurately evaluate their own people with perspective to discern what is keeping them from being a high-performing team. Beyond an accurate evaluation, leaders need to take certain actions to make the proper teamwork behaviors happen. A common pitfall is assuming that understanding the team concept equates to creating a team that works effectively.

Experience with operational teams, at nuclear power plants and in general industry, has shown that there are eight prominent themes important to assessing SLT effectiveness (table 11–1). These areas are not unique to nuclear teams but apply to other production facility leadership teams with high reliability and safety drivers as well. These are the areas that operational leadership groups need to master to become high-performing teams. "Horizontal alignment," for example, refers to how well team members work cross-functionally, as seen in their level of coordination and in how proactively they communicate with one another. Daily interactions should reflect cooperation and effective communication. Team members need to demonstrate alignment in their level of urgency in order to be able to rely on one another when addressing plant issues or people problems. They also work to achieve concurrence on when and how to address important, longer-term issues. Accountability can be an asset or, in some cases, a detriment, depending on whether team members differ on the degree to which they hold their people accountable for delivering results and for demonstrating proper behaviors. When department leaders become consistent in building accountability through their organizations, site results are seen in performance.

Table 11–1. Operational senior leadership team assessment areas

Assessment Area	Description/How Results Are Demonstrated
Horizontal Alignment	Degree and quality of cross-functional coordination and cooperation are high; SLT members communicate proactively; they work together to solve problems daily; results are seen in work management effectiveness.
Vision, Mission, and Priorities	SLT members are in agreement on station vision, mission, and priorities—how to achieve station goals daily and in the long term. SLT has regular conversations about priorities and long-term vision; results are seen in the rate of overall plant improvement.
Expectations—Roles and Responsibilities	Expectations regarding roles and responsibilities (RR) are clear and understood; RR expectations include teamwork behavior; SLT members are consistent in their execution of RR; results are seen in clean hand-offs in daily work and ownership demonstrated by each organization.
Processes and Structure Support Teamwork	SLT makes opportunities to communicate regularly and accomplish the business of running the operation; meetings, schedules, and shared accountabilities are structured and executed to foster cross-functional coordination and challenge; results are seen in meeting effectiveness, with the right people at the meetings.
Standards	SLT members share a pursuit for higher standards, measure gaps to excellence, and continually strive to improve; SLT members are self-critical; results are seen in effective use of quality and oversight organizations and the corrective action program.
Urgency	SLT members show a common, appropriate level of urgency to address issues and obtain safe results; one department's urgent concern is a concern for all; SLT members work toward alignment on what is urgent and important and when and how to address longer term important issues; results are seen in material condition of the station.
Accountability	SLT members are consistent in holding individuals and organizations accountable; performance issues are addressed in a timely manner for all departments so the broader organization strives for high levels of performance; results are seen in safety and production indicators.
Workforce Engagement	SLT members are collaborative in their efforts to involve employees in improvements; communications are consistent, frequent, and include input from employees; results are seen in personnel safety and use of corrective action program.

Team Development Example

Using the eight key assessment areas for operational teams listed in table 11–1, there are ways to assess and intervene with teams to help make them successful. DTE Energy's Fermi 2 station provides a good case study to illustrate how the analysis of, and interventions with, the SLT can have a significant impact on site performance. An effectiveness review was conducted two years after the initial assessment and subsequent to a number of team interventions at the station.[4]

DTE Energy, based in Detroit, has a single nuclear generating facility in operation—Fermi Unit 2—in Newport, Michigan. Fermi 2's generation capacity, industrial safety record, and other indicators had started to show decline in 2006 after a few years of good performance. In 2007 DTE named a new chief nuclear officer, Jack Davis, and site vice president, Joe Plona, who set out to turn around station performance by addressing organizational concerns (in addition to equipment issues), starting with the senior leadership team.

The SLT effectiveness work at Fermi was part of a larger organizational change effort to make significant and sustained behavioral changes and performance improvements.[5] The interventions with the site SLT were important first steps in getting them completely aligned on what was needed to turn around the broader organization.

The SLT assessment process

The SLT assessment included individual interviews, a cross-functional team survey, on-site observations, and review of archival information on the site and team. The assessment at Fermi included all the key managers and directors (15 people) from various departments who were responsible for the daily operation of the plant. Together they have accountability for addressing the longer term concerns of the station (e.g., equipment reliability).

The semi-structured individual interviews with team members targeted the eight key areas of operational team performance in addition to open-ended questions about the team and station. A second part of the assessment included evaluating the individual leadership capabilities of the team members. How well do the skills and leadership behaviors of each team member match the needs of the current environment? Poor-performing teammates who struggle in their current position can be a significant drag on a team's development and ability to lead a turnaround.

In order to identify specific cross-functional team behaviors, a brief survey was used. The cross-functional team survey asked participants to rate the team on statements such as, "We challenge one another in meetings to arrive at the best answers," and "When we need help from another department, we get it immediately." The second part of the survey asked participants to force rank the departments on their cross-functional teamwork from best to worst. The cross-functional team survey was given in 2007 prior to the start of team development work and again in 2009 as part of the effectiveness review.

During observations of team performance, assessors looked for demonstrations of team behaviors tied to the eight key areas as well as cross-functional or other behaviors potentially beneficial or problematic for team effectiveness. Archival data, such as past INPO evaluation reports and Nuclear Safety Review Board (NSRB) comments, were reviewed to identify themes and historical issues relevant to current team performance.

SLT assessment results

The team assessment revealed a number of underlying issues that contributed to the gaps in teamwork that needed to be addressed. Senior team members were not aligned on level of urgency, expectations regarding raising standards, and increasing accountability to improve performance. People tended to under-respond to missed deadlines, poor work products, and performance problems. The SLT was not leading improvements across the station in a coordinated manner. Instead, they were hunkered down, focused on their own departments, and reluctant to jump in and challenge or help one another. The current structures and processes did not adequately support cross-functional coordination at the SLT level. That is, there were not enough regular meetings using a format that fostered communication for the people who run the station on a daily basis. Such standard meetings are important for information sharing, problem solving, and team development.

Each team member received an individual feedback session on leadership and team behaviors that emerged from the interviews and observations. Individual leadership capabilities and potential coaching opportunities were reviewed with senior leadership as well.

Cross-functional team survey results. The scores on the survey showed that there were some cross-functional team behaviors that were much more problematic than others. As can be seen in figure 11–1, those behaviors rated the lowest were:

- When we need help from another department, we get it immediately.
- Departments share information and communicate with one another in a timely manner.
- When someone is being a poor team player, he or she receives feedback and is held accountable for it.

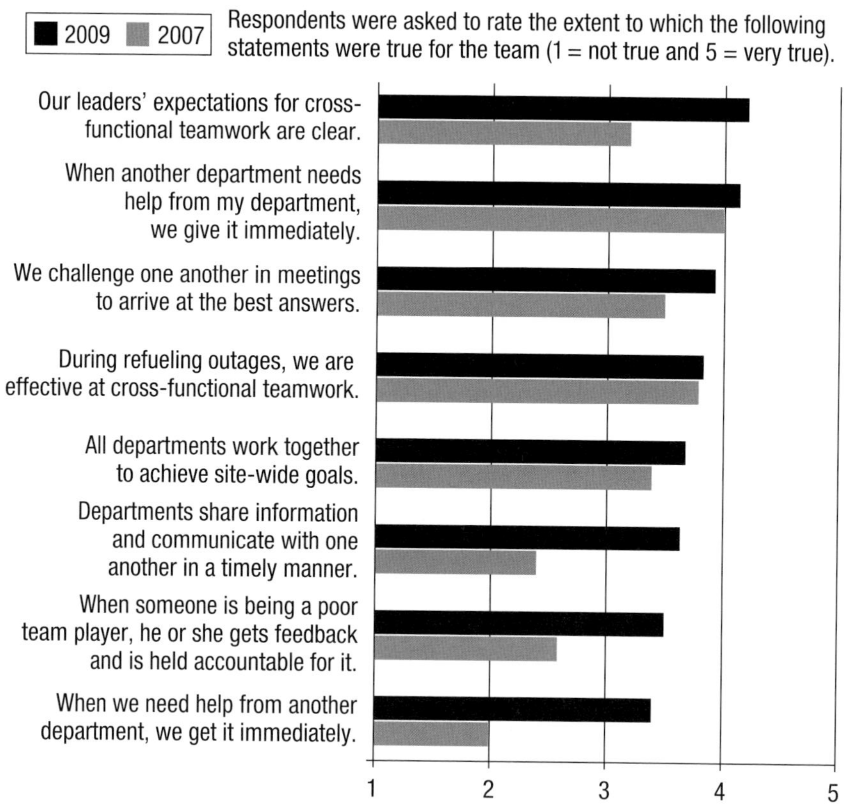

Fig. 11–1. Cross-functional team effectiveness survey, 2007 vs. 2009

Note that the statement, "When another department needs help, we give it immediately," was rated high, while "When we need help, we get it immediately," was the lowest rated item. These findings as well as the department rankings on teamwork revealed some trends and issues that were used in facilitated team sessions discussed below.

SLT interventions

After reviewing the results of the team assessment, a series of interventions were planned. Teamwork sessions were designed to build horizontal alignment on standards, urgency, and accountability, and generally to work together better to achieve results. Each session had a written agenda with objectives that were discussed in advance with top leaders. The team sessions progressed from smaller to larger circles of senior leadership, starting with small groups of top leaders, while the latter sessions included the entire SLT. Specifically, the first team session solidified alignment among the top leaders, the CNO, site VP, and plant manager, followed by a team session with the CNO, site VP, plant manager, and two top directors, before team sessions were held with the entire SLT.

Team-building session example. The first full SLT session included the top 15 managers from across the departments as well as the site VP and plant manager. The objectives of the session were to clarify expectations regarding teamwork behavior, to identify gaps in cross-functional team behavior, and to build agreement and commitment on improving teamwork. Senior leaders made initial remarks, tying the team development work with the site vision and improvement plans. Following a discussion of ground rules, the core of the team activities in this session involved going through the results of the teamwork survey.

Seeing the results of the departmental forced ranking on team behavior stimulated earnest and challenging discussion of why groups were ranked low or high. Team members were struck by the impact of their own actions and how their peers saw their behavior. Subsequently, the team reviewed the lower and higher ratings on team behavior and discussed which items were more important. The team made individual commitments regarding behaviors that affected team functioning. In sum, this session was seen as pivotal in aligning senior managers on team behavior and in obtaining commitment to one another to improve cross-functional cooperation. Additional team sessions were conducted to continue to fortify team functioning in particular areas.

Organizational support. To provide ongoing support for teamwork and alignment efforts and to create more forums for addressing issues as an SLT, the following organizational adaptations were made:

- Weekly management alignment meetings were held with the senior team to keep team initiatives active, to keep the expected team results in front of people, and for team members to hold one another accountable for behavior change.
- Organizational change and teamwork behavior initiatives were integrated into the site business plan; action plans were created and progress was reviewed regularly by the leadership team.
- Training was conducted to reinforce skills related to communication and mutual accountability.

Teamwork effectiveness review

In 2009, the team and the station had made noticeable progress. The SLT reported having shared expectations for performance improvements and better cross-functional teamwork. The site had improved operational performance, including greater capacity factor and a better industrial safety record. The effectiveness review was conducted to validate organizational improvements as well as to identify new areas that needed attention.

The cross-functional team survey that was taken in 2007 was readministered in order to compare results on cross-functional behavior. At this point, the SLT also took the Strategic Talent Solutions (STS) Team Audit, a 36-item questionnaire that provides a more in-depth assessment of team alignment, capability, cohesion, and communication.

Team members' perceptions of cross-functional teamwork were much more positive in the 2009 survey. Most improvement was seen in the following items:

- When we need help from another department, we get it immediately.
- Departments share information and communicate with one another in a timely manner.

- Our leaders' expectations for cross-functional teamwork are clear.
- When someone is being a poor team player, he or she receives feedback and is held accountable for it.

The STS team audit showed strong alignment and capability across the team. Team members reported that they were clear on the strategy, focused on getting results, supportive of the team's goals, and understood what they were accountable for delivering. The area of cohesion was the lowest-rated area, especially collegiality and support. Follow-up team sessions were held focusing on cohesion in order to continue to build the resilience of the SLT.

Conclusions

Fermi made good use of the team assessment and intervention work. Getting alignment from the top and expanding team sessions methodically helped reinforce results in subsequent team sessions. Rigorous follow-up on recommended actions was critical as the site made the oversight of team behavior part of its ongoing processes. The effectiveness review is a useful tool to validate improvements, highlight new areas for growth, and ensure that team development resources are used wisely.

Translating Nuclear SLT Lessons to Other Industries

Nuclear organizations generally accept the fact that the effectiveness of the senior leadership team is critical to leading the site to high performance, although they do not always have or use the tools to create and sustain a high-performing SLT. Sites that sustain high performance tend to proactively manage the makeup and functioning of the SLT. High-performing teams also have structures in place that support the team. This includes having regular team meetings with the right participants and an agenda that fosters teamwork, putting team behavioral expectations into performance plans, and including team goals in site business plans.

Leaders and general managers in heavy industry continually balance production and safety demands in a challenging environment. They also benefit from using teams with complementary roles and intentionally working on team behaviors. The following are suggested initial steps to build a strong team that gets results:

1. Bring a cross-discipline senior team together to discuss overall operational goals and how they are expected to work together as a team to reach the goals. Define who is on the team and why.
2. Set up a schedule that allows the same team to spend time together on a regular basis to share information, solve problems, get aligned on issues, and develop as a team.
3. Distribute and discuss written roles and responsibilities based on input from team members. Work through challenging conversations discussing real examples to build alignment on responsibilities and ownership.
4. Do not underestimate the degree to which problem team members can slow down the team and organization from making progress. Address poor performers and poor team players individually, and build a team of capable individuals.
5. Spend time getting team members in sync on standards, accountability, urgency, and workforce engagement. Use real plant examples and work through differences together.
6. Regularly check and adjust. Intermittently ask for objective and external feedback on the team and team members—avoid overreliance on a single vantage point.

References

1 Rogers, Mary J., and Colleen M. Walker. 2010. "Evaluation of Leadership Teamwork Interventions at DTE Energy—Fermi 2 Nuclear Power Station." *Energy Workforce*, Summer. http://strattalent.com/~strattal/sites/default/files/Eval_Lead_Tmwrk_Pennwell%20Energy%20Workforce%20-%20DTE%20Fermi%20Senior%20Team%20-%20VF%20-%202010-02-08.pdf.

2 Katzenbach, Jon R., and Douglas K. Smith. 1993. *The Wisdom of Teams*. New York: HarperCollins.

3 Institute of Nuclear Power Operations. 2007. *Leadership Fundamentals to Achieve and Sustain Excellent Station Performance*. Atlanta, GA: INPO. September.

4 Rogers, Mary J., and Colleen M. Walker. 2010.

5 Rogers, Mary J., and Ben K. Fearing. 2009. "Organizational Effectiveness Review: Leadership Assessment Gap Analysis for Fermi 2." *Strategic Talent Solutions*, July.

APPENDIX A Talent Development Checklist

The best nuclear power operators apply the same rigor and oversight to their management and leadership development programs as they do to their technical programs, recognizing the importance of talent development to the success and sustainability of the station. The talent development checklist can be used for nuclear as well as nonnuclear facilities and companies as a self-assessment tool that identifies potential gaps in the program's implementation.

The talent development checklist (table A–1) includes a wide range of processes and actions that exemplify a thorough and effective talent development program. Further, it includes an indication of how well each process or action is demonstrated or documented. For example, if a site has a formal process for succession planning, is it a fully written procedure with complete instructions, easily accessible to people who need to use it? Or, in the other extreme, is it just talked about and partially written or found only on the human resource manager's computer? The checklist includes a rating of the degree of formality and rigor as well as specifics on how the process or behavior can be verified through any type of documentation or demonstration. Use of the checklist can provide detailed feedback on gaps in talent development processes and behaviors, as well as in the formality, structure, and documentation supporting the program.

Table A–1. Talent development checklist

For each item, note the type of documentation or demonstration (e.g., self-report, observation, written policy or process, or formal procedure) and rate the robustness of the documentation/demonstration: minimal or absent, modest, or high caliber.

	Robustness of the Documentation/Demonstration		
Item	Minimal or Absent	Modest	High Caliber
The site has documented processes for succession planning and for management and leadership development.			
The site follows its documented processes for succession planning and development.			
Line leaders provide oversight to succession and development programs and processes to ensure they are executed, effective, and prioritized adequately.			
Leaders serve as role models in the energy they spend developing their own succession candidates and coaching and mentoring others.			
Leadership development and training programs are seen as effective by participants.			
Leadership roles are seen as desirable opportunities by succession candidates.			
Line leader and HR roles and responsibilities as they pertain to succession planning and leadership development are clear and executed accordingly.			

Item	Robustness of the Documentation/Demonstration Minimal or Absent	 Modest	 High Caliber
HR personnel dedicated to management and leadership development are capable and viewed as highly effective by line leaders.			
Assessments of potential succession candidates are conducted by trained professionals.			
Decisions about leadership changes and promotions are made methodically, with adequate input from appropriate parties.			
Effectiveness reviews are conducted regularly and focus on process, behavior, and results.			
Other:			
Other:			

APPENDIX B

Cross-Functional Teamwork Questionnaire

The cross-functional teamwork questionnaire is a simple tool to assess specific teamwork behavior, such as proactive communication, cross-discipline support, and working together on common goals, that are important for overall operational effectiveness. The questionnaire also includes items pertaining to expectations and accountability and a forced ranking of departments on cross-functional teamwork.

This questionnaire can be amended to use with specific teams. For example, the overall directions and the instructions for the ranking of teamwork can be edited to focus participants on a particular team under consideration. The results can be used to highlight areas of weakness in cross-functional teamwork and as a conversation starter for team building meetings.

For each item, please indicate the extent to which you agree with the following statements as they pertain to your team.	Strongly Agree	Agree	Neither Agree nor Disagree	Disagree	Strongly Disagree
1. Different departments share information and communicate with one another in a timely manner.	○	○	○	○	○
2. When my department needs help from another department, we get it quickly.	○	○	○	○	○
3. We challenge one another in meetings to arrive at the best answer.	○	○	○	○	○
4. When someone is being a poor team player, he or she gets feedback and is held accountable.	○	○	○	○	○
5. During refueling outages, we are effective at cross-functional teamwork.	○	○	○	○	○
6. All departments work together to achieve site-wide goals.	○	○	○	○	○
7. Our leaders' expectations for cross-functional teamwork are clear.	○	○	○	○	○
8. When other departments need help from my department, they get it quickly.	○	○	○	○	○

9. Rank each department in the organization in the order of the best cross-functional team players to the poorest cross-functional team players, with 1 being the best.

   1.
   2.
   3.
   4.
   5.
   6.
   7.
   8.

Fig. B–1. Cross-functional teamwork questionnaire

APPENDIX C Nuclear Power and Related Organizations

The nuclear industry's cooperative culture is represented in numerous industry groups. These groups meet throughout the year on various technical and nontechnical topics, and members share information, including solutions to problems, and offer extensive networking opportunities. Other groups provide social networking and informal discussion groups for interested parties. Table C–1 includes a list of nuclear and other related organizations, basic information on the purpose of the group, and contact information.

Table C–1. Nuclear power and related organizations

Organization	Description	Contact Information
American Association of Blacks in Energy (AABE)	National organization of energy professionals with 40 local chapters; dedicated to ensuring the input of African Americans and other minorities into energy policy.	1625 K Street NW, Ste. 405 Washington, DC 20006 (202) 371-9530 info@aabe.org
American Nuclear Society (ANS)	Large association of nuclear energy professionals with multiple publications, working groups, and conferences.	555 North Kensington Avenue La Grange Park, IL 60526 (708) 579-8287 ans.org
Boiling Water Reactors Owners' Group (BWROG)	Sponsored by GE; membership is limited to employees, customers, and suppliers; purpose is to partner with customers to improve reliability and share resources.	site.ge-energy.com/BWROG
Facebook networking groups	Nuclear Power Plants This Week in Nuclear	facebook.com

Organization	Description	Contact Information
Institute of Nuclear Power Operations (INPO)	Membership is limited to nuclear power operators, suppliers, their employees, and INPO employees; mission is to promote the highest levels of safety and reliability in commercial nuclear power plants.	700 Galleria Parkway SE, Ste. 100 Atlanta, GA 30339 INPO.info/
International Atomic Energy Association (IAEA)	World's central intergovernmental forum for scientific and technical cooperation in nuclear energy; works to promote global nuclear security.	Headquarters in Vienna. Liaison office at the United Nations: 1 United Nations Plaza Room DC-1-1155 New York, NY 10017 USA (212) 963-6010 iaeany@un.org iaea.org
LinkedIn professional networking groups	Energy and Utilities Network Nuclear Power-The Next Generation Nuclear Safety Networking for Nuclear Engineers and Nuclear Professionals	linkedin.com
North American Young Generation in Nuclear (NA-YGN)	Members are nuclear professionals age 35 or younger or within the first 10 years of their career; older professionals may join as honorary members; purpose is to promote professional skills, inform the public, and network.	na-ygn.org
Nuclear Energy Institute	Membership is open to companies, governments, unions, and universities. NEI is the policy organization of the nuclear energy industry; objective is to ensure the formation of policies that promote the beneficial uses of nuclear energy and technologies in the United States and around the world.	1776 I Street NW, Ste. 400 Washington, DC 20006 (202) 739-8000 nei.org

Organization	Description	Contact Information
Nukeworker.com	Web site provides breadth of information for nuclear power workers: nuclear power news, job postings, union contact information for traveling craft workers, and a discussion forum for nuclear power–related topics.	nukeworker.com
Pressurized Water Reactor Owners' Group (PWROG)	Sponsored by Westinghouse; membership is limited to employees, customers, and suppliers; purpose is to improve the competitiveness of member plants through cost-effective resolution of issues common to more than one member, and to provide a superior regulatory interface in support of member activities.	wog.westinghousenuclear.com
STARs Alliance	Membership is made up of PWR nuclear power single-site utility owners; purpose is to collaborate and share best practices among the members in order to continually improve performance.	starsalliance.com
U.S. Energy Information Administration	Collects, analyzes, and shares independent and impartial energy information to promote sound policy making, efficient markets, and public understanding of energy.	eia.gov

Organization	Description	Contact Information
U.S. Nuclear Regulatory Commission (NRC)	A U.S. independent governmental agency created by the U.S. Congress in 1974 to regulate commercial nuclear power plants and nuclear materials; purpose is to ensure the safe use of radioactive materials for beneficial civilian purposes while protecting people and the environment.	U.S. Nuclear Regulatory Commission Washington, DC 20555 (800) 368-5642 (301) 415-7000 nrc.gov
U.S. Women in Nuclear (WIN) and WIN Global	Established to support an environment in nuclear energy and nuclear technologies in which women and men are able to succeed. Provides a network for professional development and to inform the public about nuclear technologies.	NEI WIN coordinator: (202) 739-8050 winus.org win-global.org
Utilities Service Alliance (USA)	A nonprofit organization made up of single nuclear site utilities; purpose is to facilitate collaboration among member utilities, improve safety and performance, and reduce operating and maintenance costs.	9200 Indian Creek Pkwy, Ste. 201 Overland Park, KS 66210 (913) 451-5641 usainc.org

Index

D

E

F

G

H

I

J

K

R

T

U

V

W